T0146046

The Skeleton
REVEALED

The Skeleton REVEALED

An Illustrated Tour of the Vertebrates

STEVE HUSKEY

Johns Hopkins University Press | Baltimore

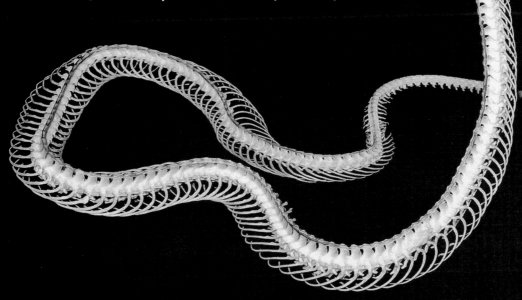

Johns Hopkins University Press
2715 North Charles Street
Baltimore, Maryland 21218-4363
www.press.jhu.edu

Library of Congress cataloging data is
available.

ISBN-13: 978-1-4214-2148-3
ISBN-10: 1-4214-2148-8

A catalog record for this book is available
from the British Library.

*Special discounts are available for bulk
purchases of this book. For more information,
please contact Special Sales at 410-516-6936
or specialsales@press.jhu.edu.*

Johns Hopkins University Press uses
environmentally friendly book materials,
including recycled text paper that is
composed of at least 30 percent post-
consumer waste, whenever possible.

For dad, you're missed every day
For Jenny, I'm grateful every day
For Emma and Ella, you amaze me every day

■ Acknowledgments

I am tremendously indebted to the many people and businesses that have been willing to donate animal carcasses to this cause over the years, including Jim Harrison and Kristen Wiley, Kentucky Reptile Zoo; Mitch and Diana Gibbs, Fishey Business; Pacific Aqua Farms; Aquascape Online; Tennessee Aquarium; The Moleman, Tom Schmidt; Art Westwood; and Kenny Barnett, among others. This book would not have been possible had it not been for my PhD mentor, Dr. Ralph Turingan, and his willingness to teach me the skeleton methodology. I am grateful to the countless students and contemporaries with whom I have worked on skeletons over the years, particularly Josh Johnson and Grant Stoecklin. I am also indebted to the many colleagues with whom discussions about all things bony made this an enjoyable endeavor, such as Dr. Justin Grubich, Dr. Christopher Anderson, Dr. Mark Westneat, and Dr. Nicolai Konow. I am thankful to my editors, Dr. Vince Burke and Debby Bors, for their helpful guidance on the book. I am also grateful to Western Kentucky University, the WKU Research Foundation, and the Biology Department, which provided countless resources for this work.

Finally, I must acknowledge two individuals who were instrumental to the completion of this volume. Without their vast knowledge of photo editing and their commitment to perfect, finished images, this book would have fallen short of its goals. As such, my heartfelt appreciation is given to

Tressa Tullis
and Dr. Matt Tullis.

■ Charles Darwin saw them. So did Alfred Russel Wallace. As did a few other naturalists even before them. Rock-like remnants of animals from long ago, buried in the earth, providing hints about what might have come before us and how this idea of evolution might work. In a period long before telephones, you can just imagine the excitement of a time when scientists communicated by mailing letters back and forth around the world, anxiously awaiting a reply from a colleague.

"What is it?"

"I don't know, but it's enormous! It looks like a giant leg bone!"

"So does the one I found!"

"Describe it to me, please!"

Fossils have likely been discovered by humans for millennia and may be the impetus for such creatures as medieval dragons and those found in Greek and Roman mythology. Something had to explain what these giant, bone-shaped rocks came from. And they weren't too far off—flying reptiles, giant meat-eaters, necks as long as a horse-drawn carriage, feet the size of carriage wheels. All buried in the ground as if waiting to spring back to life at a time when many still believed in the spontaneous generation of living creatures.

Fossils reveal a great deal about extinct animals, yet leave so much to the imagination. They are borne from hard plates, scutes, nails, teeth, and bones. Our estimates of age, relatedness, feeding habits, and more are predicated on relating fossils to what we see living around us today. By understanding features of present-day organisms, scientists can form hypotheses about what an extinct organism was capable of and how it lived. The most critical of these features is the vertebrate skeleton.

Internal skeletons appear to have evolved in a couple of different ways. Some skeletal elements formed from external, bone-like, protective plates that eventually became internalized. Most skeletal features arose as hard minerals became deposited around the cartilage precursors to bone as a means to make them stronger. In fact, today's animals with bony skeletons begin life with a skeleton composed entirely of cartilage. This flexible material forms the framework for future bones as the animal develops and grows.

Skeletons can function as protection, such as a skull protecting the brain inside it. They can function as a reservoir for critical elements, such as calcium and phosphorus. And they can be the site of red blood cell formation inside their marrow.

Teeth are also considered part of the skeleton because they are often firmly attached to bones and, like bones, are superbly preserved as fossils. However, teeth probably first evolved from the tiny, armored scales of ancestral shark skin known as dermal denticles. These scales eventually wrapped around early jaws and assisted in capturing and processing new types of food; the tooth race was on! Now we see teeth of every size, shape, and function lining the jaws of nearly every vertebrate that chews.

Other important skeletal features include the fin rays, spines, and barbs of the fishes. The soft rays of bony fish are essentially a series of stacked, bony scales capable of being flexed for swimming or maneuvering. Spines evolved from single dermal denticles in cartilaginous fishes and from scales or bones in other fishes. They work as protection, reproductive appendages, cut-waters, noisemakers, and flamboyant displays to attract mates or fend off rivals. Barbs, like those on the tail of stingrays, also evolved from single dermal denticles and serve as protection from threats. When a stingray is pinned, the tail is thrust upward and flexed to reveal the barb with its sharp point, serrated edges, and venom-delivering grooves.

Most importantly, skeletons can work as a system of levers on which muscles can pull to generate movement. Because movement underlies behavior, we can use our understanding of skeletons to make predictions about extinct animals such as dinosaurs. How fast did *Tyrannosaurus rex* run? What did *Apatosaurus* eat? How did *Triceratops* defend itself? By modeling *T. rex* after today's ostriches, it's apparent that they leaned forward when they walked or ran, rather than standing upright like a kangaroo, and that they could move rather quickly. By comparing the spade-like teeth of *Apatosaurus* to the incisors of other browsers, we know they were herbivores capable of snipping vegetation, and lots of it. By observing horned chameleons and beetles that use their points to battle rivals and defend themselves from predators, we can safely assume *Triceratops* did the same.

Skeletons tell us nearly everything we need to know about the vertebrates of today as well. Who were their ancestors? How big can they get? What do they eat and how do they capture their prey? Can they fly, glide, run, jump, burrow, climb, or swim? If so, what underlies their ability to perform these tasks and how has evolution shaped it for millions of years? By simply examining the skeleton of today's vertebrates, one can gather a remarkable amount of information about the species with which we share our planet. Scientists use that information to better understand how an animal's morphology (form and structure) assists it in meeting the daily challenges of survival.

Remember that every day in the wild illustrates survival of the fittest. Humans lose sight of that because we rarely experience natural selection in our technologically advanced, sheltered lives. Have you ever had to fight to the death for a meal? Would you be able to climb a tree to avoid being eaten? When was the last time you were afraid you wouldn't survive to see the next sunrise? For an animal, how its body assists it in relating to the world around it can mean the difference between finding a meal and being eaten. The coordinated actions of the vertebrate skeleton allow animals with backbones to successfully navigate their daily tasks, whether that's defending a territory, securing a mate, protecting their young, or capturing a meal. Any one of these can mean the difference between living another day or losing in the lottery of life; the vertebrate skeleton is at the interface of these two.

My book is a survey of a portion of the remarkable functional diversity found in vertebrates alive today. Their carcasses have been positioned in life-like poses, processed by flesh-eating beetles, and meticulously reconstructed to reveal how their skeletons help them relate to the outside world and accomplish tasks. There is a vast array of amazing things the vertebrate skeleton makes possible. The evolution of the vertebrate skeleton, as revealed in these pages, is still a source of wonder, something Darwin and Wallace knew, and something I hope you will discover as well.

■ *Morelia spilota*, the carpet python, is a constrictor found in Australia, Indonesia, and New Guinea. These pythons inhabit a vast array of habitats, from cool, wet rainforests to hot, arid islands. Reaching nearly 13 feet in length, they are usually the top predator in places where they occur. They eat small- to medium-sized mammals, bats, and birds and, as juveniles, often eat lizards. A nocturnal species, they are equally as comfortable in trees as they are on the ground, using tree limbs to snatch perched, sleeping birds or bats that fly too closely. Six rows of rear-facing teeth hold their prey, four rows on top and two below. Prey is quickly wrapped in a tight coil, and death occurs rapidly.

Carpet pythons are popular in the snake trade because of their beautiful skin coloration and relatively docile demeanor as adults. The species does not appear to be at risk of overharvesting from the wild, though habitat degradation and competition for space with an ever-growing human population is always a threat.

■ The gulf flounder, *Paralichthys albigutta*, is a flatfish found from North Carolina to Texas in salt and brackish water, from inches to 200 feet deep. They spend most of their time on sandy bottoms where their camouflage blends in perfectly. They also use a wiggling motion to stir up sand along their edges that settles back down on top of these fish, helping to conceal them.

Gulf flounder begin life like any other larval fish—bilaterally symmetrical, with eyes on each side of the head. However, very early in their lives, the right eye migrates over the top of the skull and takes up residence on the left side. At the same time, the fish turns to create an eyeless, blind side that lies on the ocean floor and an eyed side that watches above. These fish have small gut cavities and completely lack a swim bladder. This makes swimming along the bottom easier, and there they pursue their shrimp, fish, and small crab prey, which they capture with large canine teeth. This species can grow to 18 inches in length and is a prized sport fish because of the fight they produce when hooked. Their wide, flat bodies and large muscle ratios create struggles that can last an hour in deep water.

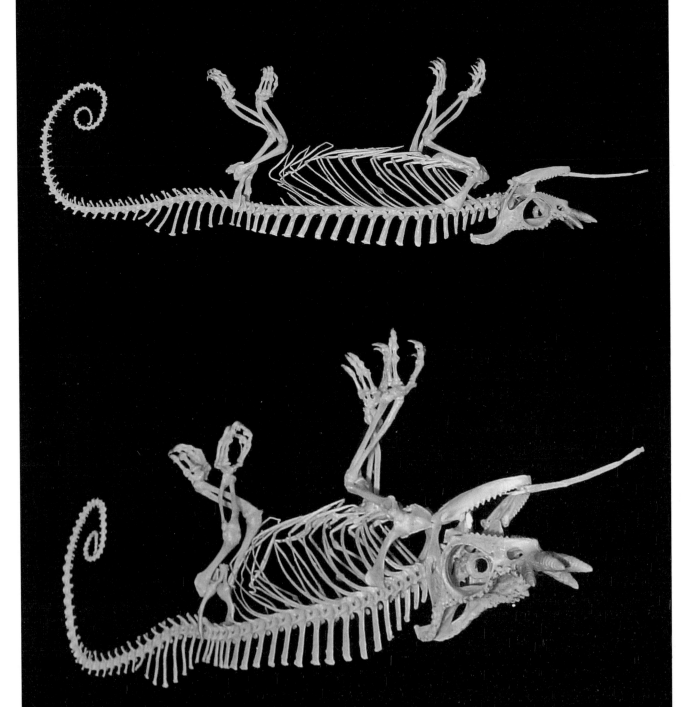

Male four-horned chameleons, *Trioceros quadricornis*, have 1 to 3 pairs of horns on their snout that point up and slightly to the sides, though two pairs is the norm. Females lack these horns. The species grows up to 9 inches long and demonstrates the amazing color-change behavior for which chameleons are known. They can display white, green, blue, yellow, orange, red, brown, and black patterns and hues to perfectly match their surroundings.

Four-horned chameleons hail from Cameroon and Nigeria. They inhabit cool, sloping cloud forests up to elevations of 6,000 feet. As such, they are capable of dealing with weather conditions that vary from nearly 32°F to 70°F where humidity levels are as high as 90%.

Like other chameleons, they appear to be a hodge-podge of anatomical novelties. Their eyes can work independently, looking in two different directions at once. Their feet have two digits on one side and three on the other side, making them perfectly adept at gripping branches. Their tails, adapted for grasping ("prehensile"), function like a fifth leg. Their tongues are ballistic missiles that shoot at their food and stick by "grabbing" prey with the tip of the tongue, while the middle of the tongue creates a suction cup to hold their meal, made up mainly of insects.

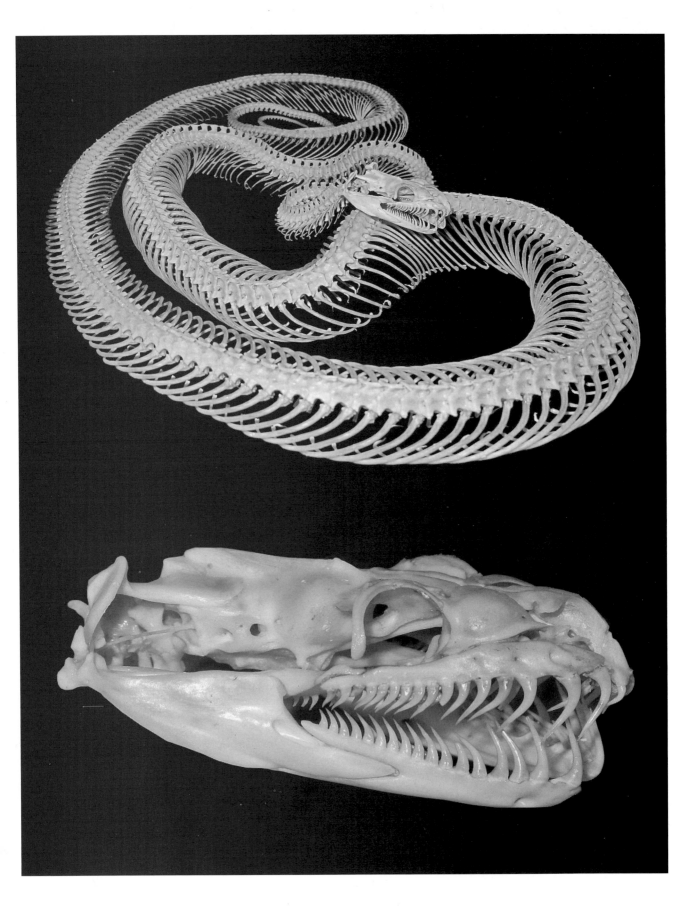

The Cooper's hawk, *Accipiter cooperii*, has one of the broadest distributions of any raptor, from Canada to northern South America. They are most common in woodland habitats, but they also use open lots to spot prey with their forward-facing eyes, which provide excellent depth perception.

Cooper's hawks are relatively small raptors, similar in size to crows. They are extremely acrobatic flyers with fast wing-beats. This allows them to maneuver through woodlands in pursuit of prey such as smaller birds, squirrels, chipmunks, and mice. They snatch their prey with powerful, sharp talons and tear off pieces of flesh with their strong, arched beak.

This individual is posed with a recently captured southern flying squirrel, *Glaucomys volans*, which it has carried back to a perch for eating. Flying squirrels are nocturnal but sometimes venture out during crepuscular periods (dawn and dusk) to locate food sources. That's when this Cooper's hawk seized the opportunity for an evening snack.

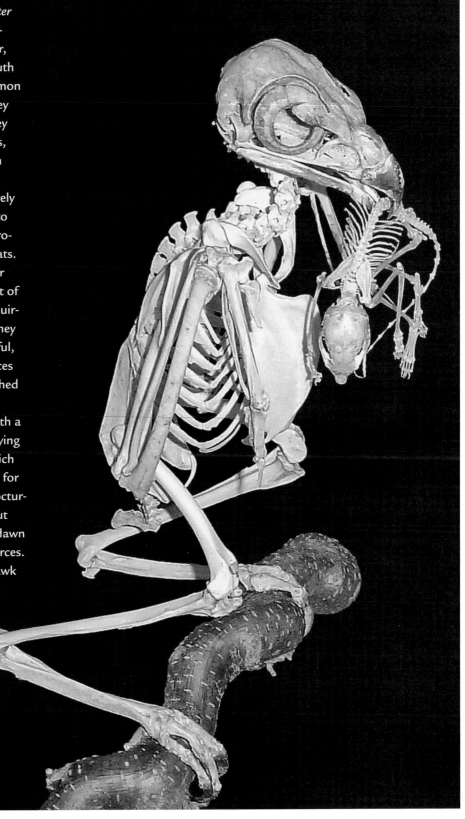

■ *Epibulus insidiator*. The name sounds like it's something special—and it is. This is the slingjaw wrasse, a marine fish common on reefs throughout the Indo-Pacific. This species has the remarkable ability to protrude its jaws further than any other animal. While it looks like many other fishes that can barely open their small mouths, the slingjaw wrasse has an uncanny ability to shoot its jaws off its face toward its prey.

The jaw hinge in this species resides much further back than in other fish this size. This hinge, along with a sliding upper jaw, allows the fish to protrude its jaws into the water at the moment of attack. This closes the distance between them and their prey and also serves to create a perfect, straw-like opening through which they can suck in their food.

Slingjaws feed on elusive reef creatures such as small crabs, shrimps, and fishes. These prey often hide near the opening of a reef crevice where they can dart in to safety if threatened. The slingjaw can swim by nonchalantly as if nothing is amiss, and then suddenly discharge its mouth toward the unsuspecting prey before they have time to escape.

■

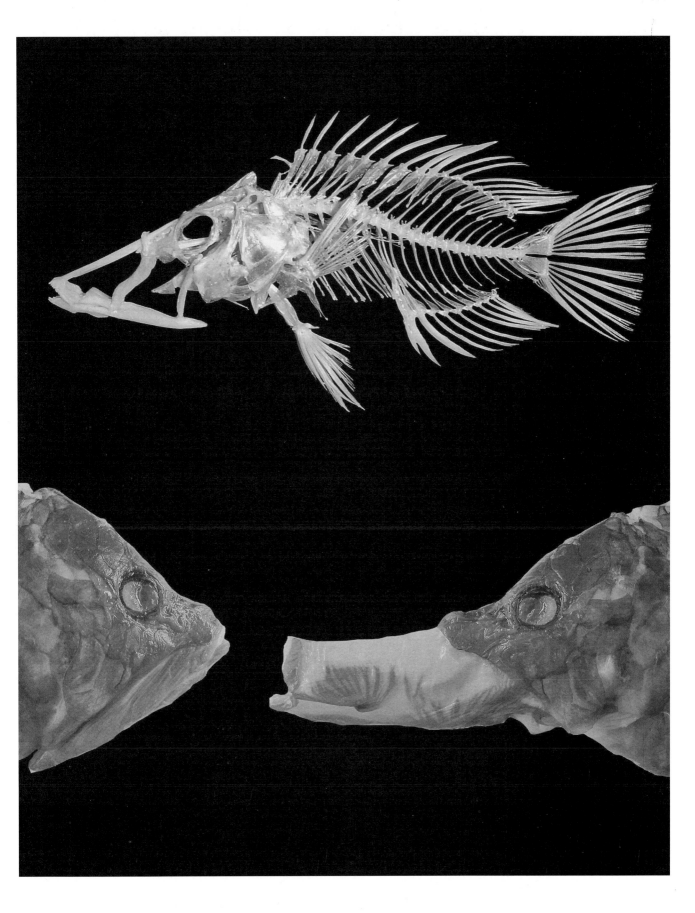

■ *Morelia viridis*, the green tree python, is a constrictor found in Australia, Indonesia, and Papua New Guinea. They inhabit rainforests and are almost always found in trees, shrubs, or bushes where their vibrant green coloration helps them hide. Reaching only 6 feet in length, they eat mainly small- to medium-sized mammals and lizards that they snatch from their hiding place among the leaves. Prey is held by six rows of rear-facing teeth, four rows on top and two below. Death comes quickly as the prey is tightly squeezed.

Lacking limbs, as all snakes do, green tree pythons have the difficult task of swallowing prey, usually bigger than their own head, without the aid of limbs to push food down their throat. Rather, they have evolved a sophisticated set of jaws, where the two rows of teeth on the roof of the mouth lie on bones that can move front to back. When they begin to swallow their prey whole, these bones and teeth effectively "walk" the skull over the top of their meal, while the lower jaws do the same from below.

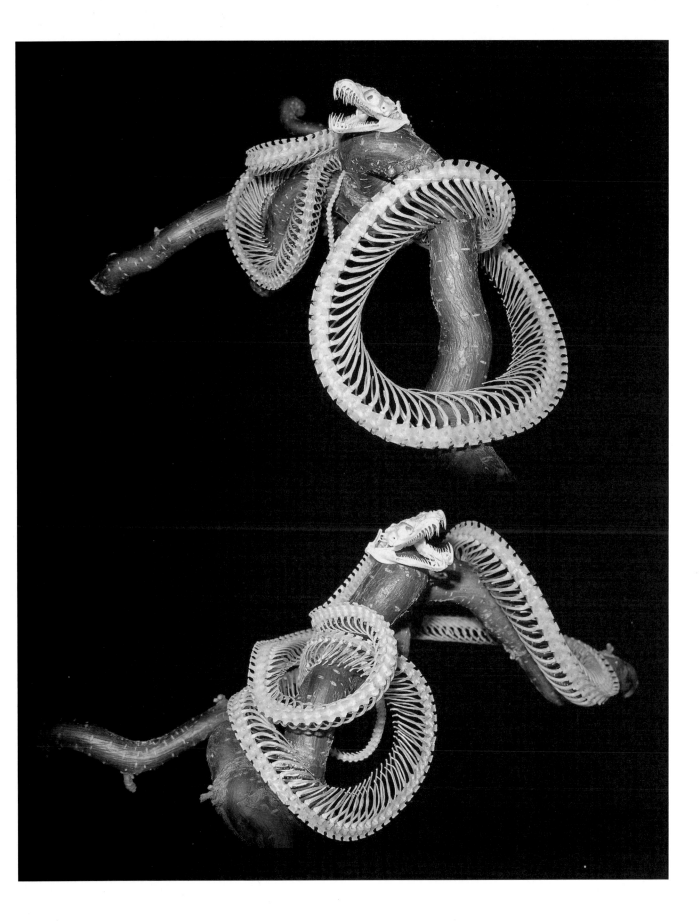

■ The Colombian lancehead, *Bothrops colombiensis*, is an easily agitated species of venomous snake found throughout eastern South America in lowland tropical rainforests. They grow to 5 feet and have some of the longest fangs, almost 1 inch in length. It is a common species encountered in forest-cleared plantations and, as such, bites on humans are not uncommon. The venom is fast acting and hemotoxic, meaning that it destroys red blood cells, and can be lethal if untreated.

This species is a perfect example of an ambush predator. They hunt during the daytime and lie in wait under leaves for hours waiting for small mammals, lizards, or frogs to venture nearby. Colombian lanceheads have a fast strike; the venom is delivered quickly and the snake backs off, waiting for the venom to do its job of killing and partially digesting the prey. The snake then tracks it down by its scent trail and swallows it whole.

■

■ *Varanus salvadorii*, or the crocodile monitor, is found in tropical forests and swamps in Papua New Guinea. They hold the title as the longest lizards alive today. Komodo dragons get much heavier, but the extremely long tail of croc monitors helps them exceed 10 feet in length. The tail functions as a whip when they are threatened and can easily split skin, possibly even breaking bones. It is also useful for balance, since this species spends most of its time in trees and has long, sharp claws to aid in climbing.

Crocodile monitors eat rodents, frogs, and eggs by swallowing them whole. They also consume carrion and are able to tear off pieces of meat with their numerous razor-sharp teeth and powerful jaws. Like other monitor lizards, crocs are known to have slightly toxic saliva that functions as venom when delivered into their prey. It's not fatal to humans, but it can result in excruciating pain, uncontrolled bleeding, and elevated blood pressure.

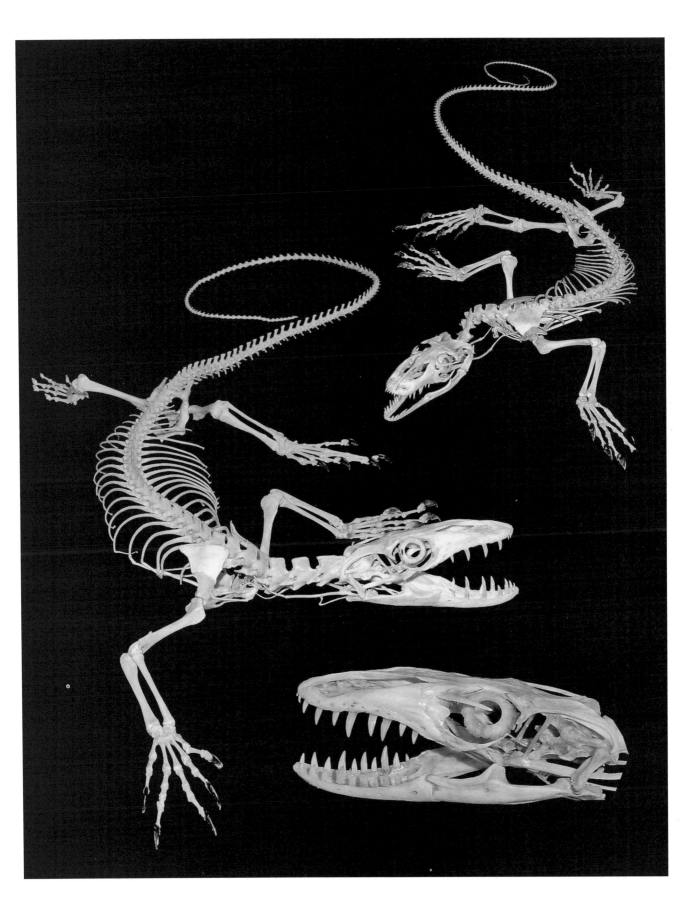

The cubera snapper, *Lutjanus cyanopterus*, is the largest of the snapper species, exceeding 125 pounds and growing longer than 4 feet. They are found on reefs and rocky ledges in waters up to 150 feet deep within the western Atlantic Ocean.

A prized sport fish, cubera snappers are tremendous fighters and will often dive into reef structures to resist being reeled in. The fight they offer, plus their mild, white flesh, makes them one of the oceans tastiest and most sought-after species. As a target in spear fishing, cuberas are wary and require stealth and long breath-holds in order to inch close enough for a shot.

Cubera snappers eat just about whatever they like. They have a mouth large enough to accommodate even the widest fish prey and jaws strong enough to process crabs and spiny lobsters. Their large canine teeth sink deep into anything they catch and leave little opportunity for escape.

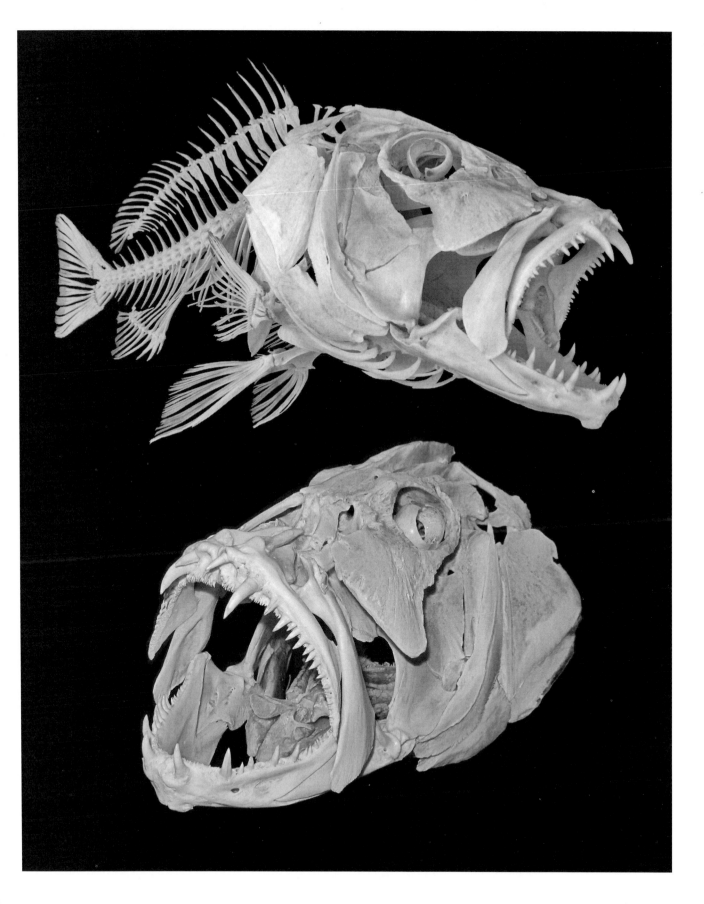

The dwarf crocodile, *Osteolaemus tetraspis* is the smallest of all the crocodiles, rarely reaching 6 feet in length. It is found in the flooded rainforests and plains of West Africa. It tends to inhabit slow-moving rivers and swamps but spends a significant amount of time in burrows it digs on land, especially during the heat of the day.

Dwarf crocodiles are nocturnal and, like all crocodilians, they are carnivores that eat whatever meat they can find, including small mammals, birds, fish, turtles, snakes, frogs, crustaceans, and carrion. Due to their small size, they are susceptible to predation by larger crocodile species, although humans are the only true threat to their existence. To help them survive predation, they are wrapped in a shield of bony plates called osteoderms. This species even has plates on their underbelly, unlike most other crocodiles. Their genus and species names literally translate into "bony throat" and "four shields," reflecting their overly armored body.

The largest of all the rattle-snakes, eastern diamondbacks, *Crotalus adamanteus*, can grow up to 8 feet long and weigh over 10 pounds. They are found on the coastal plains of the southeastern United States in habitats that range from dry, sandy areas and dunes to palmetto stands and hardwood forests.

Eastern diamondback rat-tlesnakes are prolific hunters, lying in wait for unsuspecting mammals to venture too close. Like all snakes, they are deaf, but they can sense vibrations of the ground created by large threats, such as humans, which initiates the high-speed activity of the rattle. Highly specialized, hollow scales comprise the rattle, which creates the clatter for which the group is known. The rattle is meant as a warning that you're getting too close to the strike zone.

Eastern diamondback venom contains toxins that affect blood coagulants, cell walls, and mus-cle cells and can be fatal if not treated immediately. They employ their venom most often to feed on rabbits and squirrels. The bottom image is of an eastern diamondback that had swallowed a grey squirrel, *Sciurus carolinensis*.

The tarpon, *Megalops atlanticus*, or silver king, as it's known by angling enthusiasts, is one of the most revered sport fish on the planet. Found along temperate, subtropical, and tropical coasts of the Atlantic, they can endure a broad range of salinities. Juvenile tarpon live in freshwater creeks that drain into estuaries, and as they grow into adults, they are found in a variety of habitats, including mangroves, coral reefs, seagrass meadows, bays, and the open ocean.

Tarpon are huge fish, exceeding 7 feet in length and weighing almost 300 pounds. Their large eyes are an adaptation for feeding in poorly lit environments, and their up-turned mouth is evolution's answer to feeding at the surface. Tarpon will pursue prey from below and launch powerful upward blasts coupled with a "roll" at the surface to capture anything that will fit in their mouth. Their huge mouth functions like a giant fish-trap when it explodes open, generating tremendous suction pressure. When coupled with fast, forward thrusts of the body, tarpon sit near the top of the food chain where they occur.

When hooked by an angler, tarpon go on impressive runs, often coupled with relentless jumps, tail-walking, and audible headshaking from the clapping of their large gill plates. Their ancient, bony mouths are difficult to hook, hence the technique of lowering one's rod when they jump, known as "bowing to the silver king," to keep them from getting enough leverage to throw the hook.

Nature's slowest mammal, the two-toed sloth, *Choloepus hoffmanni*, is found in the tropical rainforests of Central and South America. They live high in the tree canopy, venturing to the ground only to relieve themselves once a week or to move to a new tree for food. They are herbivores, feeding mainly on leaves, shoots, fruits, nuts, flowers, and sap, though they have been known to eat bird eggs, insects, and small lizards. Their teeth are quite sharp and are used for slicing and grinding vegetation.

Since sloths spend their entire lives upside down, they have some interesting adaptations. First, their fur parts along their belly and runs toward their back, so rainwater sheds more easily. Second, their body temperature varies greatly depending on the environment, which is a rare trait in mammals. Third, they move at only about 6 to 9 feet per minute, which makes them so slow that algae actually grow on their fur. In fact, they have special grooves in their fur to house the algae, because it turns their brown fur green, which helps hide them in the trees.

Sloths look fluffy and fat, as they are only about 25% muscle, which is about half that of most mammals. But most of their body size is the result of their thick coat. They do not need lots of muscle, because they spend most of their days hanging upside down using their long, strong, hook-like claws. Because of their slow speeds on the ground or in the trees, they are preyed upon by anacondas, caiman, harpy eagles, jaguars, and ocelots.

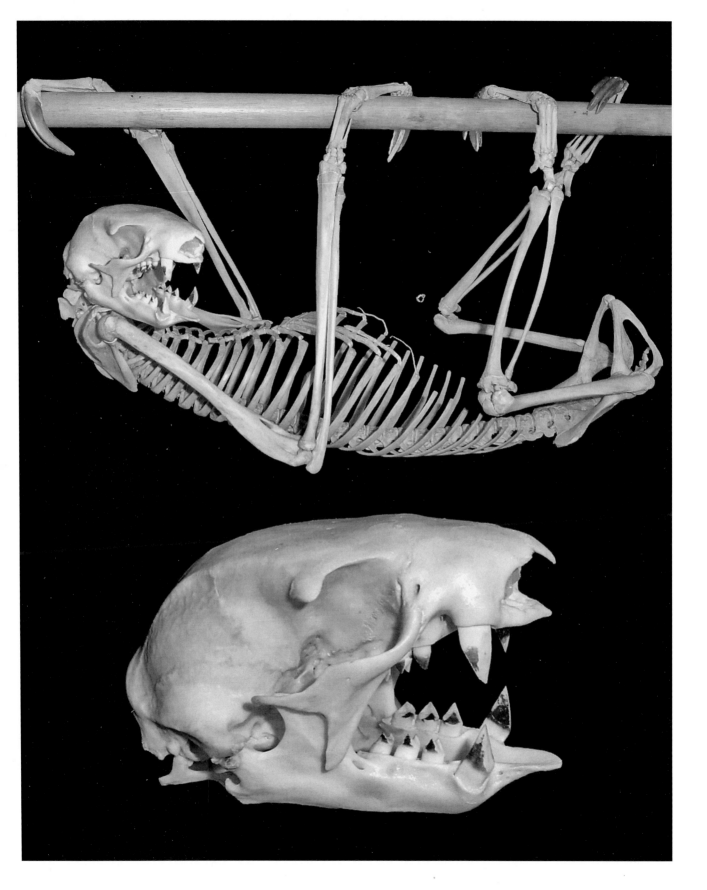

The aptly named pancake batfish, *Halieutichthys aculeatus*, is a member of the bottom-dwelling family known as Ogcocephalidae, or "big-heads." Its genus name means "the fisher that fishes," because this is a fish that fishes for other fishes.

They inhabit sandy ocean floors as deep as 2,500 feet, from the southern United States throughout the Gulf of Mexico, where they bury themselves in the sand and lie in wait for unsuspecting fishes or shrimps to venture too close. The "fisher" part of their name comes from the

specialized spine that projects off their face, just above the mouth. They can twitch this spine and, with the soft fleshy appendage at its tip, attract prey close enough to be snatched up by some of the fastest jaws in the world.

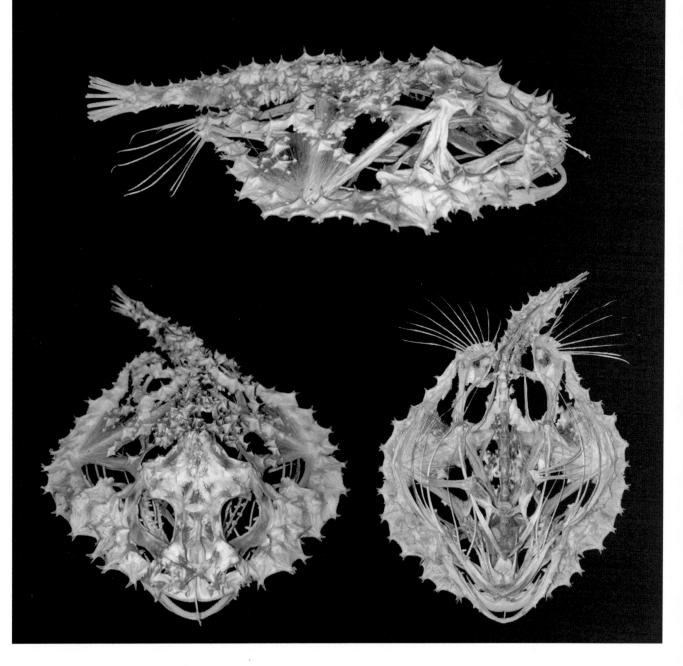

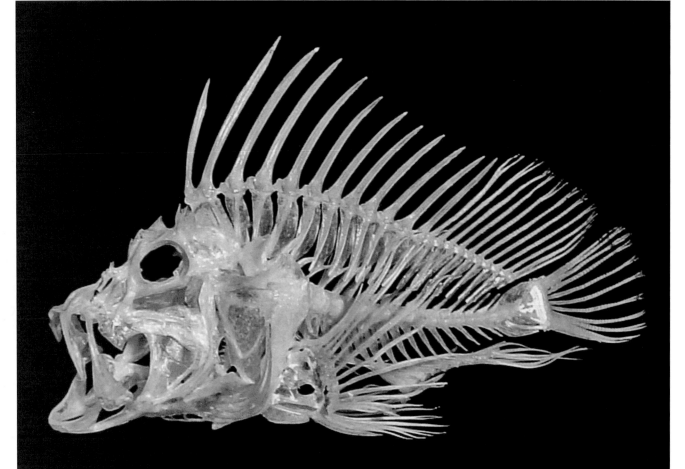

The leaf scorpionfish, or paperfish, *Taenianotus triacanthus*, gets its name from its appearance and its behavior. Its flattened body resembles a dead leaf lying in the water, and the fish will even sway from side to side to look like a dead leaf tossed by a wave. They do this to slowly approach unwary prey and to avoid drawing the unwanted attention of potential predators.

Leaf scorpions use their pectoral fins, positioned low on the body, to "creep" along the reef in slow motion to sneak up on unsuspecting prey. They eat smaller fish species, as well as shrimps, and do so with one of the fastest sets of jaws in the sea. Their mouth opens with such enthusiasm that it creates tremendous suction pressure to draw in prey, and this happens many times faster than the human eye can blink.

Leaf scorpionfish are widespread on coral reefs in water up to 400 feet deep. They occur from the East African coast, through the Indo-Pacific, to Hawaii. Their color variations include green, brown, red, pink, yellow, and white and usually match their surroundings very closely. Like other scorpionfishes, they are venomous and introduce the venom through their dorsal and anal fin spines. The venom is only slightly toxic to humans, however, and a sting can be treated for a quick recovery.

The largest of all the true cobras, the forest cobra, *Naja melanoleuca*, can grow up to 10 feet long. The species occurs in Central and Western Africa and typically inhabits lowland forests and moist savannas, with occasional sightings in drier habitats. The forest cobra is also a capable swimmer and will patrol water edges in search of unsuspecting prey.

The prey forest cobras feed on is variable, depending on the habitat they spend the most time patrolling. Those in wet habitats will readily eat fish, while grassland forest cobras eat almost exclusively small mammals. They are also great climbers and will pursue other reptiles, such as lizards and snakes, into trees.

Forest cobras are members of the family Elapidae. Like other elapids, they have fixed fangs, which means they do not swing on "hinges" like those of snakes such as vipers and rattlers. Since their fangs do not fold back when not in use, they must be much shorter than mobile-fanged snakes. A 9-foot forest cobra may only have one-quarter-inch fangs. However, those fangs deliver copious amounts of neurotoxic (destructive of nerve cells or nervous tissue) venom. A bite will yield drowsiness, paralysis, fever, and extremely low blood pressure and can be fatal due to neurological and respiratory failure if left untreated.

The hood of a cobra is created by ribs that are modified to be much straighter than the rest. When threatened, the snake will rear up, flare its hood, and hiss as a warning to stay back. This species is considered very aggressive and will rush forward to strike if cornered. As such, they are often persecuted in their homeland, when the root of the problem is habitat loss and fragmentation by an ever-growing human population.

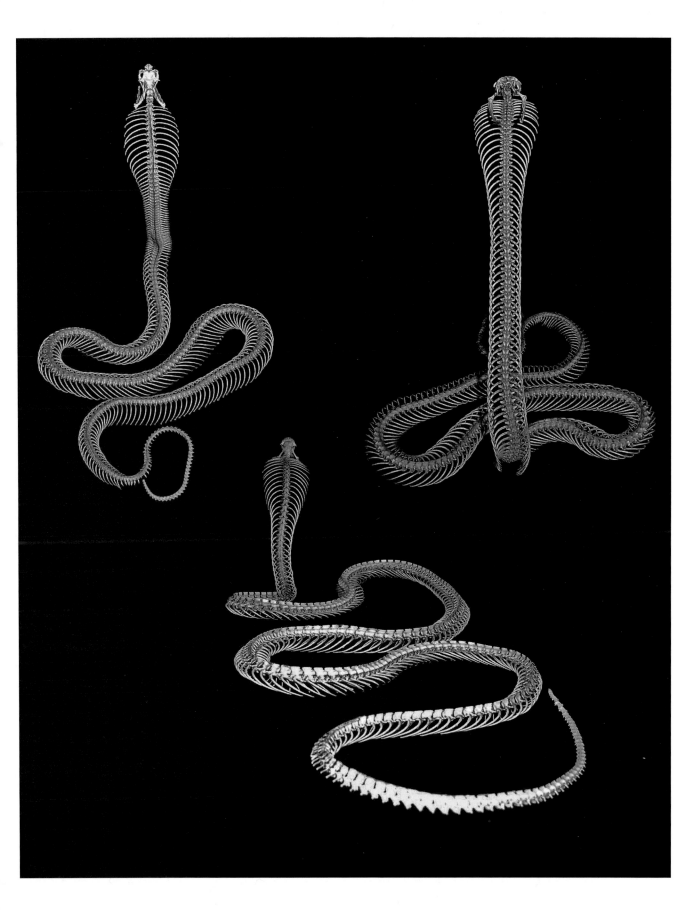

The scrawled cowfish, *Acanthostracion quadricornis*, smooth trunkfish, *Lactophrys triqueter*, and yellow boxfish, *Ostracion cubicus*, are members of tropical and subtropical marine environments around the world. The first two are found in the Atlantic Ocean, while the last is found in the Indo–West Pacific. They are slow, cumbersome fish that seem to scuttle along grass beds and/or reefs in search of benthic (found at the bottom of a body of water) prey, such as algae, shrimps, worms, bivalves, snails, and tunicates.

As members of the fish family Ostraciidae, they all have an interesting hard body covering called a cuirass that is composed of fused dermal plates. This structure is like a suit of armor and provides protection from predators. However, it also compromises their ability to flex their bodies during swimming and thereby slows them down. Therefore, most move slowly by flapping their pectoral fins, employing their tails only during faster swimming.

■ The bushmaster, *Lachesis muta*, is the longest venomous snake species in the New World, reaching nearly 10 feet in length. They inhabit Central and South American rainforests and possess some of the best leaf-litter camouflage of any snake. They use this camo to hide along travel corridors of small mammals, sometimes lying in wait for weeks at a time. Once struck, prey experience venom that has hemorrhagic, coagulant, and neurotoxic elements, and this venom is delivered by some of the longest fangs of all venomous snakes, nearly 1.5 inches.

Bushmasters are the only pit vipers known to lay eggs. The mother will guard the nest without leaving for the entire two and a half months until hatching occurs. The hatchlings then wriggle off to find their own territory and their first meal.

■

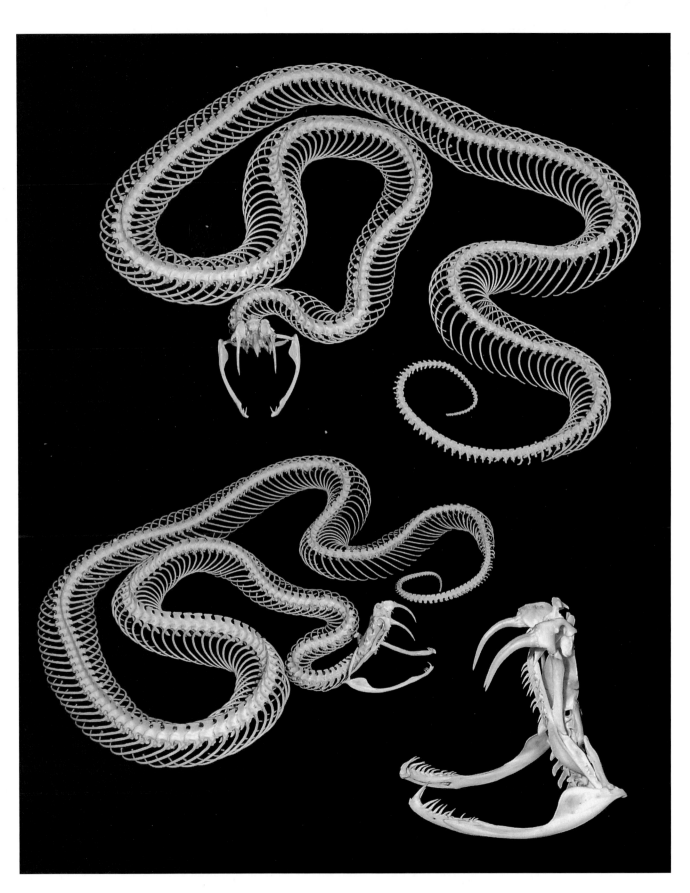

■ *Balistes capriscus*, also known as grey triggerfish, are one of the most flamboyant fish in the sea, not in their coloration but in their actions. The species inhabits reefs, worm-rock, and rocky ledges in water up to 180 feet deep. They are highly territorial and have been known to attack divers by biting their fins and sometimes their ears.

Grey triggerfish, or leatherjackets, as they're sometimes called, have some of the strongest jaws in the ocean. The species only weighs up to 14 pounds and is rarely longer than 24 inches, but they have the jaws, teeth, and bite muscles of something much larger. Their prowess at biting comes from their affinity for prey that most other fishes must ignore, such as crabs, urchins, bivalves, sand dollars, and sea stars. Greys can reverse the "suction-pump" in their mouth and use it to "spit" at their prey. They use this behavior to roll their prey over to get at the soft underbelly.

Triggerfishes get their name from the suite of three spines on their back. The first can be erected as an anchor when the fish swims into a crevice to hide from a threat. The second spine locks underneath the first, securing it in place. At this point, the spine is so secure that it will break before it gives way. However, there is a ligament that runs from the top of the third spine to the base of the second spine. If you pull on the third spine, you can disengage the second spine, and the first spine drops like the hammer on a gun—hence the name *triggerfish*.

■

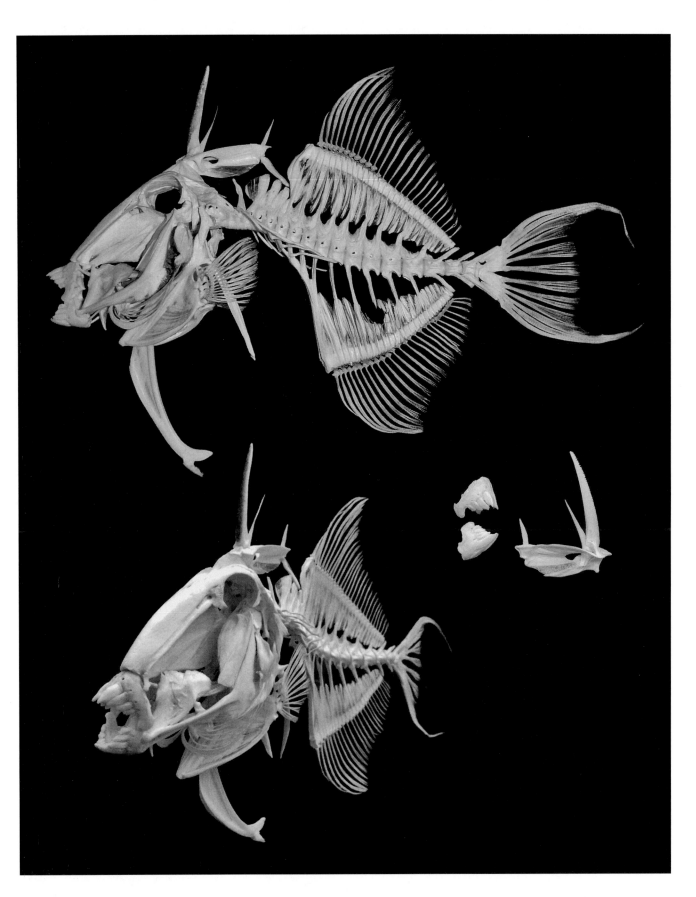

The northern caiman lizard, *Draceana guianensis*, is found in Brazil, Colombia, Ecuador, Peru, and the Guianas. As their common name implies, they are more like caiman in their daily habits than they are like their close relatives, the whiptails and tegus. Caiman lizards spend their entire day in and around water, especially in habitats such as flooded woodlands, where they can escape to the trees to avoid predators or roost at night.

Very accomplished swimmers, they have flattened tails that efficiently propel them through water in search of prey such as snails, clams, crawfish, and crabs. To eat prey this hard, caiman lizards have specialized molar-like teeth able to endure the powerful forces generated by their large jaw muscles. A morsel is positioned between the molars in the back of the jaws and is crushed. Often, the shell is removed and expelled before the soft inside is swallowed.

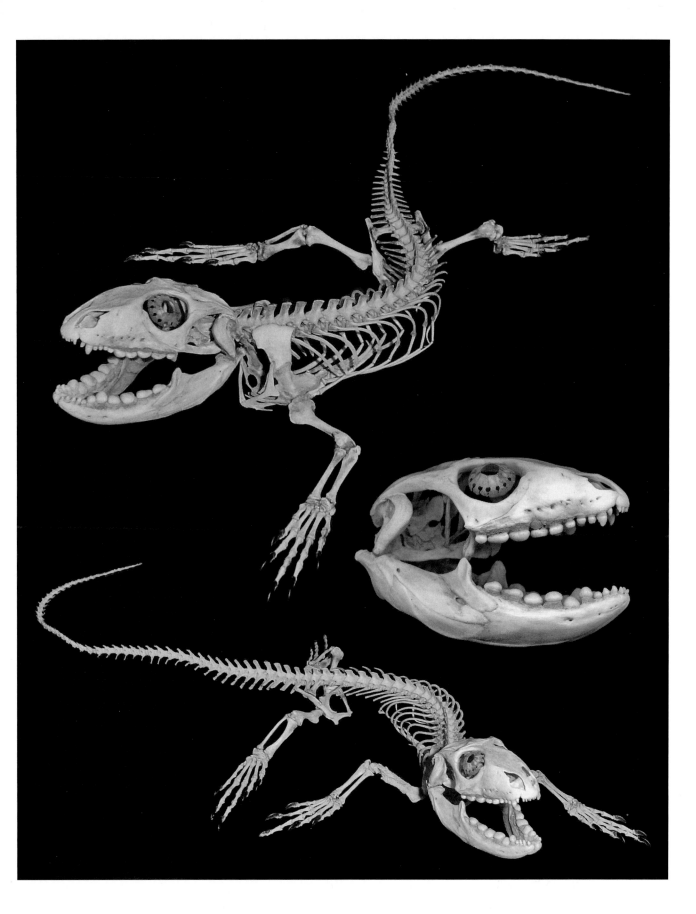

The Virginia opossum, *Didelphis virginiana*, is a common North and Central American marsupial species with whom many are familiar. Because of their tendency, especially when young, to "play 'possum" by faking death when threatened, opossums are quite easy to approach. They are also fond of human structures, especially during colder temperatures, so we encounter them in our garages, barns, and even homes.

Adult opossums are far less likely to feign death. Instead, they stand their ground, bare their large teeth, and hiss loudly. They have a rather nasty bite focused through their long, sharp canine teeth. This bite is usually used to feed on a diversity of prey, including small mammals, ground-nesting bird eggs, lizards, frogs, invertebrates, carrion, fruits, seeds, grasses, and leaves. With their short powerful front legs and large claws, they are efficient diggers. They are also quite competent climbers, using their prehensile (grasping) tail to aid in balance and grip.

As marsupials, baby opossums have to climb to their mother's pouch to develop. A female can have up to 25 babies at once and may reproduce 3 times in a year, so they are prolific breeders. Adults and babies are common prey for snakes, owls, raccoons, foxes, coyotes, bobcats, and other nocturnal predators.

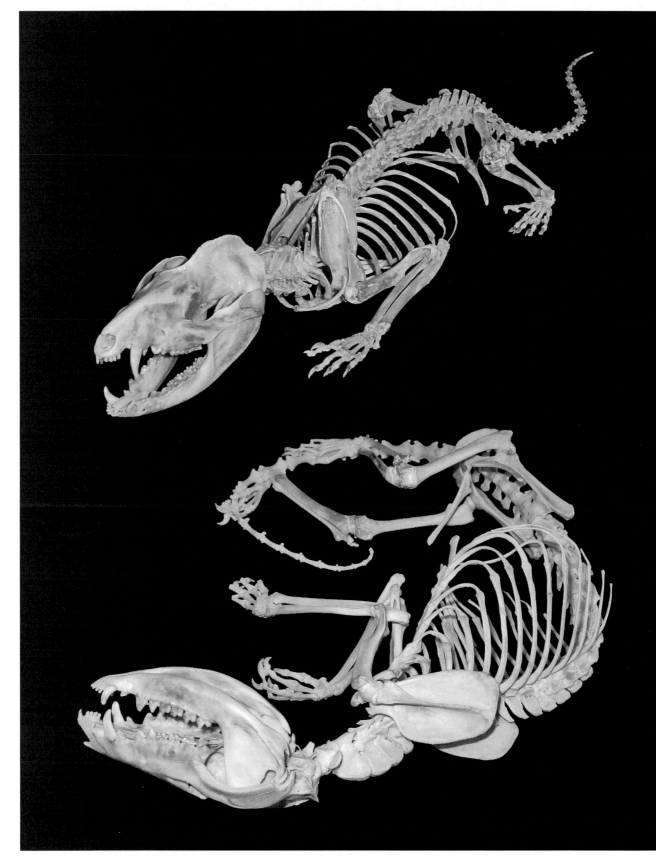

■ *Pseudelaphe flavirufa*, the Central American rat snake, called a nightwalker in Honduras, is a slender-bodied snake that can exceed 5 feet in length. It is found in coastal plains and lowlands and ventures into tropical forests in search of tree-dwelling food.

Mainly a nocturnal species, they eat small mammals, birds, amphibians, and reptiles and do so by snatching prey with fast strikes after approaching very slowly and deliberately. Prey are subdued, wrapped in a coil for control, and swallowed whole, often while still alive. This individual is posed with a green iguana, *Iguana iguana*, which it captured while hunting. Prey is swallowed head first to allow the legs or wings to fold back during the process. As with all snakes, the trachea opens far forward in the mouth so the snake can breathe while it goes through the laborious task of swallowing prey much larger than its own head.

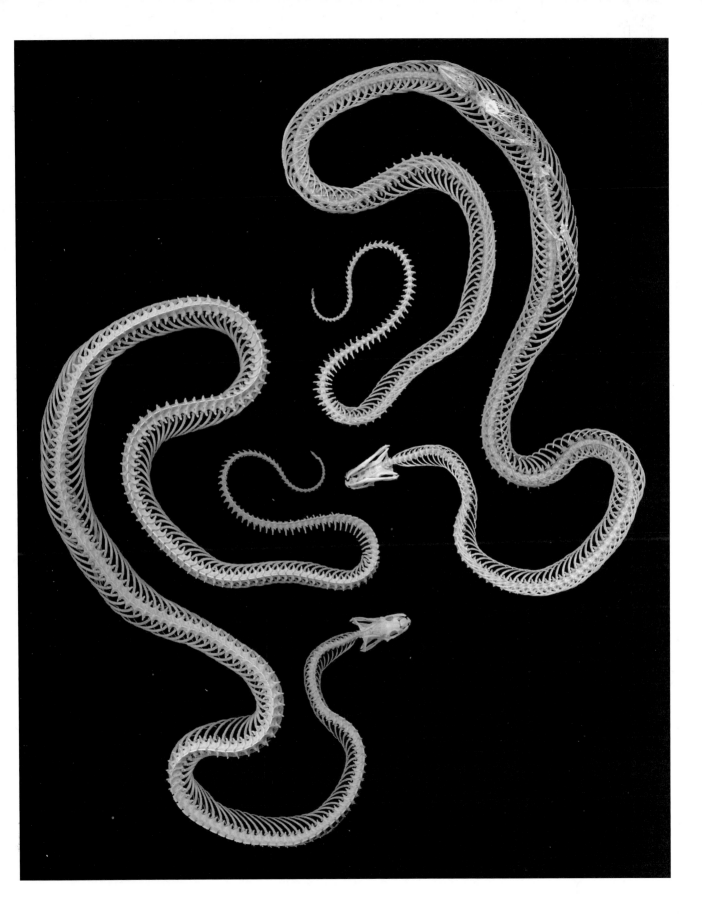

The Parson's chameleon, *Calumma parsonii*, is considered the second-longest species of these exotic lizards, at over 27 inches long, and is the heaviest chameleon species of all. It's distinguished from others by its massive, flat casque (a helmet-like protrusion) at the rear of the skull. Males have two large bony projections extending off the rostrum (snout); females, like this one, lack these bony horns.

Parson's chameleons are endemic to Madagascar and St. Marie Island. They inhabit tree canopies in ravines and canyons. Here, they hunt mainly insects, using an ambush technique of sitting on one branch for the entire day waiting for prey to venture nearby.

Like other chameleons, they are a hodge-podge of anatomical novelties. Their eyes can work independently, looking in two different directions at once. Their feet have two digits on one side and three on the other side, making them perfectly adept at gripping branches. Their prehensile (able to grasp) tail functions like a fifth leg. Their tongues are ballistic missiles that are shot at their food and stick by "grabbing" prey with the tip of the tongue, while the middle of the tongue creates a suction cup to hold their meal.

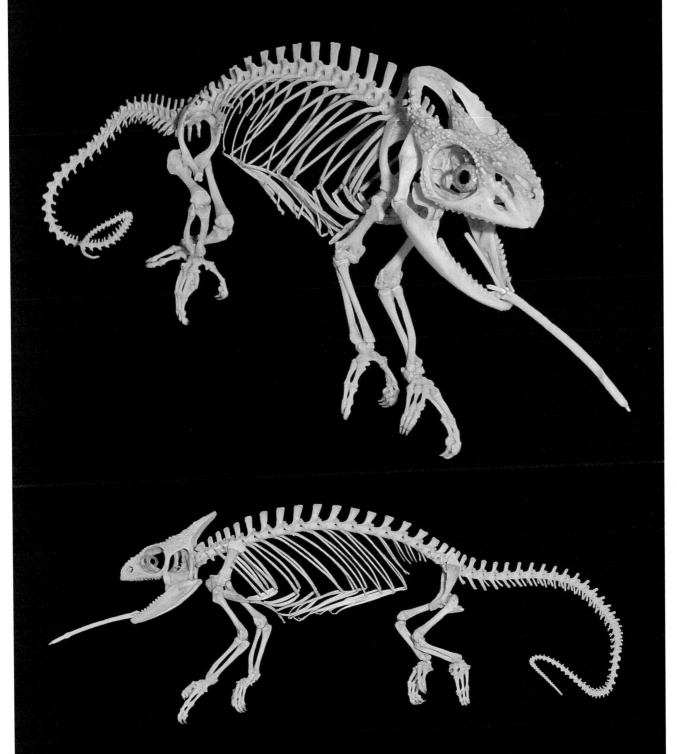

■ Koi, *Cyprinus carpio*, is a variety of the common carp bred for ornamental ponds and water gardens. They originate from Asia and Europe but have been introduced to every continent except Antarctica. Cultivation of this species as a food fish dates back to the fifth century BC in China. Koi were selectively bred for certain colors and patterns as long as one thousand years ago and can be found in white, black, red, yellow, blue, and cream.

Koi kept by aquarists and pond enthusiasts are prized for their colors, patterns, and scalation. There are currently over twenty-five named varieties of koi, and some individuals can sell for thousands of dollars to collectors. More than any other culture, the people of Japan value koi; the word *koi* is a Japanese homophone for "love," so koi are often viewed as symbols of affection and friendship.

Koi rarely exceed 3 feet long or weigh more than 25 pounds, but wild *Cyprinus carpio* have been caught at much larger sizes, up to 6 feet long. They are omnivores and will eat just about anything organic, including algae, insects, fish eggs, plant matter, and detritus (particles of dead organisms and fecal matter), which they slurp with their downward-facing mouth and fleshy lips. Their dorsal spine serves as protection and is lined with two rows of rear-facing barbs, which get buried in the flesh of any of the many animals that try to eat them, including musky, pike, bass, herons, otters, and raccoons.

The leopard tortoise, *Geochelone pardalis*, gets its name from the spotted appearance of its shell. This species is one of the largest tortoises alive today, growing to nearly 2 feet in length and weighing up to 70 pounds. It is also one of the longest-lived tortoises, reaching over 75 years of age.

Leopard tortoises are endemic to Southern and Eastern Africa from sea level up to 10,000-foot elevations. They live in grasslands and shrubby areas but need enough shade to escape the afternoon sun. They may spend much of the midday in dens created by jackals, anteaters, and other burrowing creatures.

They are herbivores (plant eaters) and get nearly all of the water they need from the grasses, succulents, mushrooms, and fallen fruits they eat. Their mouths are lined with a hard beak made of keratin that is sharp enough to slice through grass and strong enough to bite through old bones, which they eat to obtain calcium.

Lesser anteaters, *Tamandua tetradactyla*, are found in the tropical rainforests, savannas, and scrublands of Mexico, Central America, and South America. They look like the mammalian equivalent of a *Velociraptor* with those sharp nails and that one giant, long claw on each forelimb. In fact, their front claws are so long that they are forced to walk on the outside of their feet to avoid digging the claw into their own palms. They use this meat cleaver of a claw to tear into ant hills and termite mounds in search of food. Their long, flexible tongue probes for residents, lapping up thousands in a day.

The tamandua, as it's also known, reaches 3 feet in length and weighs more than 15 pounds. They would be a preferred morsel of predators such as jaguars were it not for the horrendous odor they produce when threatened. They are also happy to stand and face their rivals with claws fully ready to slash open any perceived threat.

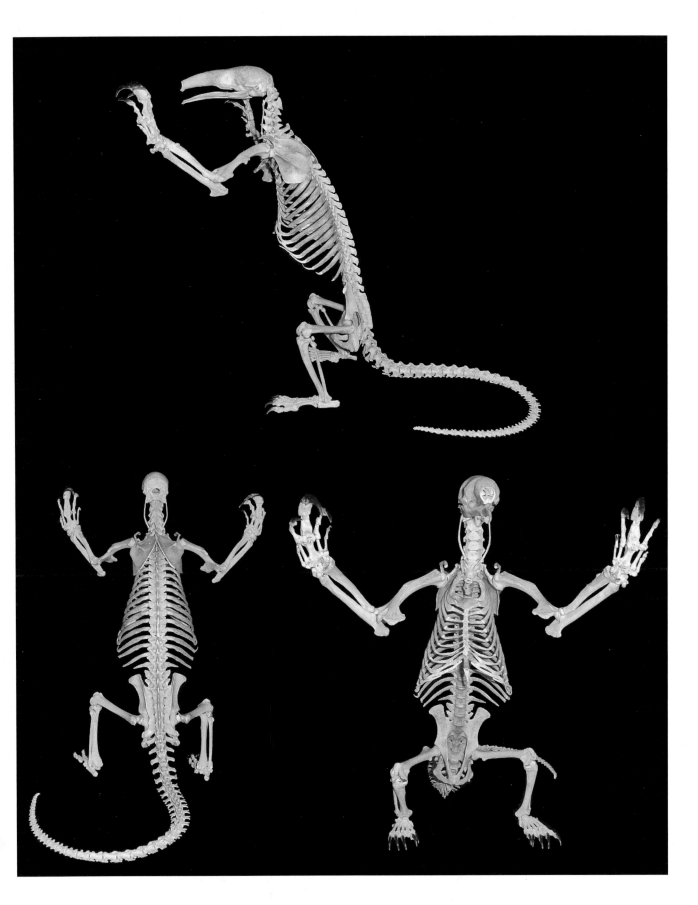

■ *Dispholidus typus*, or boom-slang, whose name translates to "tree-snake" in the Afrikaans language, is a highly venomous snake from sub-Saharan Africa. This species is almost exclusively arboreal (living in trees), venturing to the ground only to pursue escaping prey or avoid predation. Their tree-dwelling habits suit them well when feeding on their favorite prey: chameleons, lizards, tree frogs, mammals, birds, and sometimes bird eggs.

Boomslangs are rear-fanged snakes that can open their mouths nearly 180 degrees in order to get their back teeth into position to deliver venom. As such, they tend to gnash and gnaw on their prey, such as this Jackson's chameleon, *Trioceros jacksonii*. Once their food is captured, envenomated, and incapacitated, they swallow it whole in the safety of the trees. The venom is highly hemotoxic, causing internal and external bleeding and, left untreated, can be fatal in humans. However, because of its timid nature and fast escape response, boomslangs rarely interact with humans practicing common sense.

■

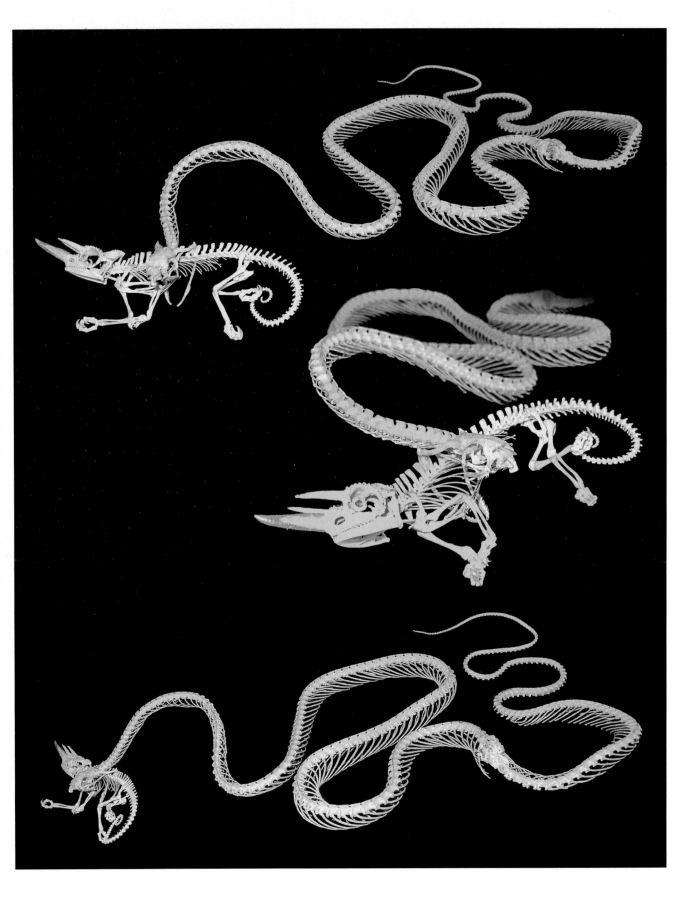

The wild turkey, *Meleagris gallopavo*, was Benjamin Franklin's first choice for the national bird of the United States. Thankfully, for those who enjoy hunting them, this was not approved—if it had been, they'd be protected from harvesting.

Wild turkeys occur throughout North America, with different subspecies existing in certain regions, such as Merriam's turkeys in the central United States, Rio Grande turkeys in the southwest, Osceola turkeys in Florida, Gould's turkeys in southern Arizona and New Mexico, eastern turkeys throughout the eastern United States, and one in southern Mexico and Central America.

Wild turkeys are large birds, weighing up to 35 pounds and standing 4 feet tall. During the spring, gobbling "toms" call to attract females. They strut by fluffing their feathers out to show off while creating a low-frequency "drum" to announce their size. When rival males meet during the spring, they often fight for access to hens. They will weave their heads together while trying to "chest-push" the other backward. When one gets the better of the other, the victor jumps on top and starts stabbing him with the long, sharp spurs on the back of his legs while grabbing his neck with his beak, as shown opposite.

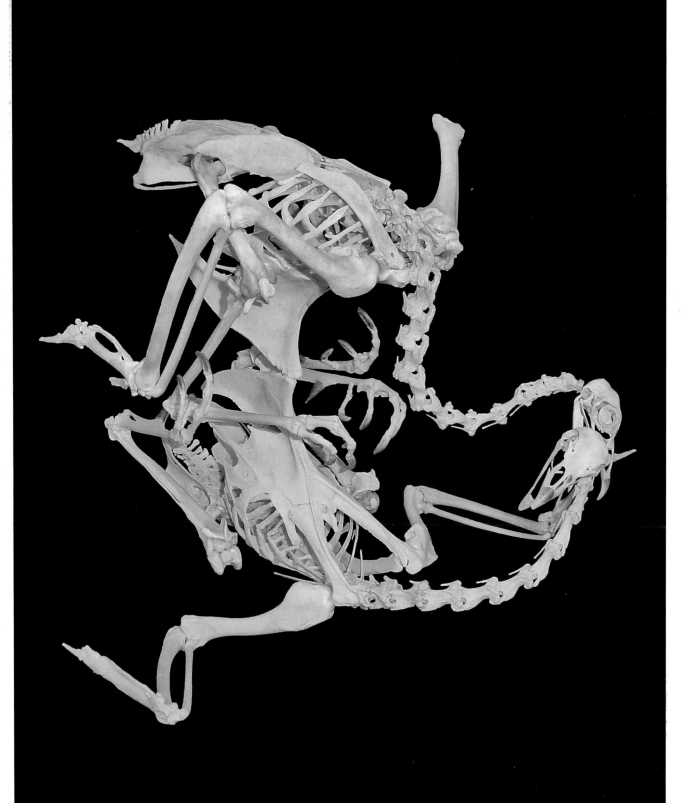

■ Gag groupers, *Mycteroperca microlepis*, are one of the most sought-after marine fish, both commercially and recreationally. Their large size, flaky white flesh, and subtle taste make sure they are pursued by everyone with a fishing pole or a spear gun. At nearly 5 feet long and as much as 80 pounds, they are one of the largest reef-dwelling, bony fish in U.S. waters.

This species occurs in the Western Atlantic from North Carolina to Brazil and throughout the Gulf of Mexico and Caribbean. They inhabit reefs and rocky outcroppings as adults and can be found on grass flats and in mangroves as juveniles.

Gag groupers have a mouth that opens as wide as their head and jaws lined with hundreds of canine teeth. These teeth are excellent for gripping struggling prey that they snatch off reef structures and out of crevices. As adults, they feed on other fishes, crabs, lobsters, squids, and octopi. As juveniles, they also eat fishes but consume significant amounts of shrimp that they find hiding on the grass flats and in the mangroves.

■

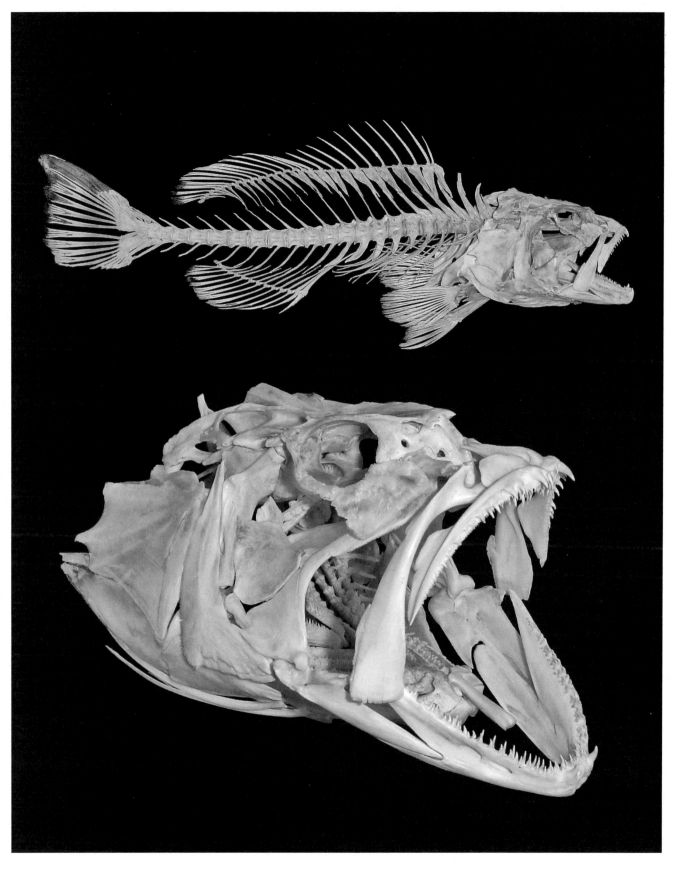

The Suriname toad, *Pipa pipa*, is actually an aquatic frog species found in the northern half of South America. They live on the bottom of muddy swamps, marshes, and rivers, surfacing only to breathe and to mate. Their flat, leaf-like appearance helps them blend in with their environment and avoid predators. It also helps them sneak up on unsuspecting prey, such as aquatic insects, worms, and crustaceans. They use star-shaped tactile organs on their front toes and a lateral line system (a system of organs that detect movement) to find prey and then launch an attack with their wide, tongueless, toothless mouth.

These frogs have one of the most interesting reproductive strategies of all animals. First, males do not croak like other frogs. Instead, they click their throat bones together to attract females. Second, during mating, the pair does somersaults through the water as they rise to the surface. The female releases eggs that get fertilized by the male and stuck to the back of the female. Then, amazingly, the eggs get embedded in the female's back as her skin grows around them. There, they develop into little froglets that hatch out of the mother's back in about three to five months.

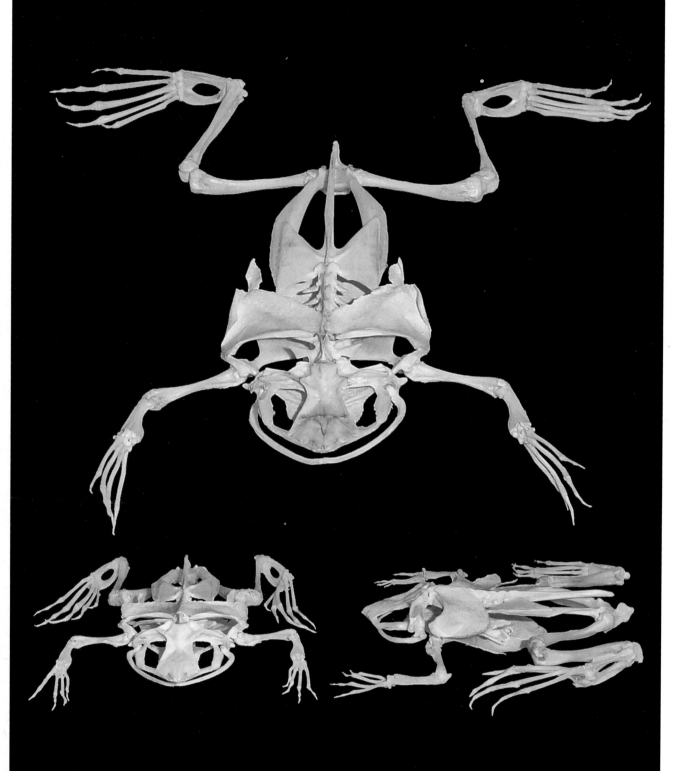

The pacu, *Colossoma macropomum*, or tambaqui in local Portuguese, is a marvel of evolutionary engineering. The species is a close relative of piranhas but differs in many significant ways. Piranhas have scalpel-sharp teeth that are used to cut through flesh and bone. Pacus are omnivores, meaning that they eat both plants and animals, with a preference for fruits and nuts that they find floating on the surface of the Amazon and Orinoco River drainages.

A pacu's mouth is lined with approximately 20 multi-functional teeth. The teeth are robust at their bases to endure powerful forces yet have sharp, cusped tips that can magnify their bite force into very high bite pressure.

When coupled with overgrown jaw muscles and thick, heavy jaw bones, the bite pressure generated by a pacu can easily crush a Brazil nut shell to get at the soft flesh inside. When corrected for body size (pacus only get up to 3 feet long and weigh 55 pounds), their bite pressure exceeds that of many large megafauna, such as bears.

A pacu's teeth are even arranged in such a manner that there are two rows of teeth on the upper jaws with a small, deep pocket between them. This allows the nut or fruit to be held in one place while the mighty lower jaw is clamped down to crush the shell.

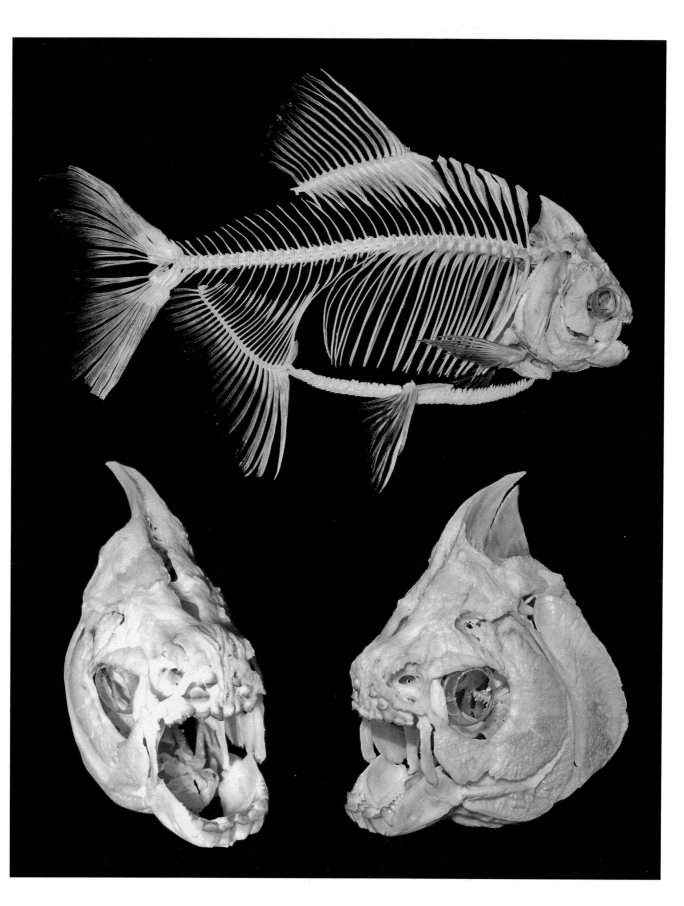

The Brazilian lancehead, *Bothrops moojeni,* is a venomous snake found in Argentina, Brazil, Bolivia, and Paraguay. They are most often found in moist, forested regions, especially those dominated by conifers. Their long fangs, nearly 1 inch in length, are used to inject venom into mammals, birds, frogs, lizards, and even some large insects.

Like other lanceheads, they get their common name from their pointed snout and slender head. As a lie-in-wait predator, they often conceal themselves under leaf litter to hide. If something too large approaches, which they can feel by detecting vibrations in the ground, they will use the tip of their tail to shake the leaves, imitating the sound of a rattle-snake in hopes of warding off any threats. Some have also been observed wiggling the tip of their tail in an attempt to lure prey close enough for a strike.

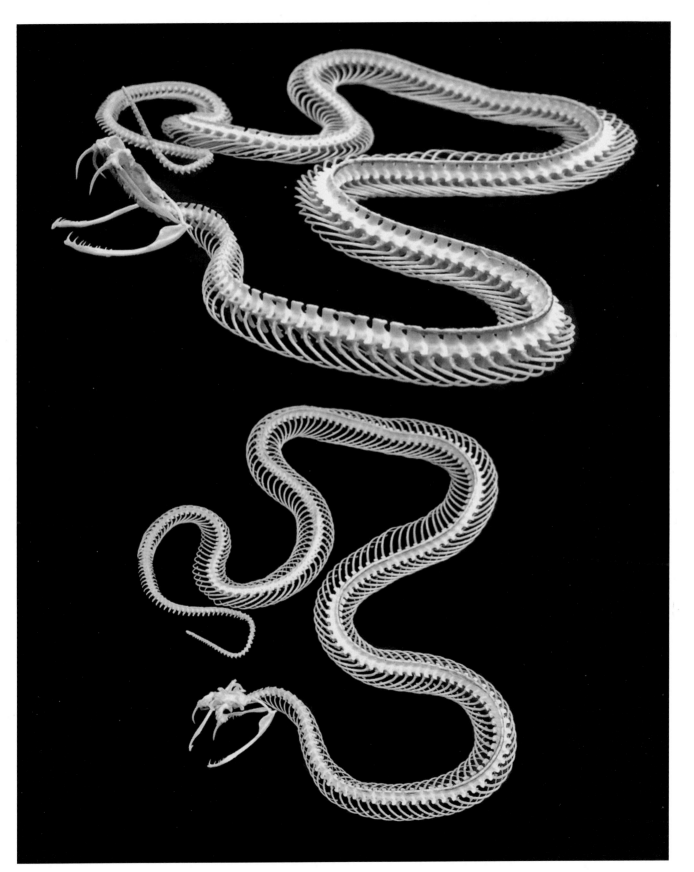

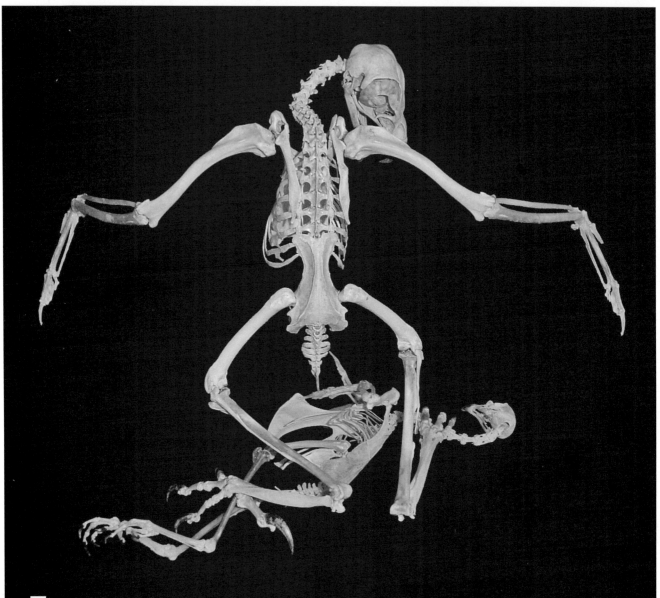

North America's most common and well-known bird of prey is the red-tailed hawk, *Buteo jamaicensis*. It is found as far north as Alaska and as far south as Mexico. Red-tailed hawks live in diverse habitats wherever they can find open country in which to hunt, surrounded by high perches from which to watch for prey.

Red-tails are known to eat just about anything they can catch, including birds as large as pheasants, mammals as large as rabbits, and even frogs, toads, insects, and snakes. Most prey are captured on the ground, though sometimes these hawks can be seen snatching birds and bats out of the sky. Smaller morsels are carried to a perch and eaten in the safety of the trees. Larger prey, such as this gray partridge, *Perdix perdix*, are slammed into, driven to the ground with talons and engulfing wings, and partially eaten on the ground before being carried to a perch.

The aptly named vampirefish, *Hydrolycus scomberoides*, or payara, is an even toothier relative of the piranhas found in the Amazon River basin. The genus name is derived from the Greek, *hydro* meaning "water" and *lykos* meaning "wolf." These "water-wolves" grow to over 1 meter long and can weigh up to 40 pounds. They are game fish prized by anglers because of their acrobatic leaps when hooked.

Vampirefish inhabit fast-moving rapids where they hide behind rocks, waiting for unsuspecting prey to struggle through the current. They dart out of hiding and swirl upward while their giant mouth is thrown wide open. This exposes the two huge, razor-sharp sabers on their lower jaw. The lower jaw is extremely long, which creates very fast speeds at its tip, allowing the fish to swing the jaw into the prey, piercing it with the sabers. Prey are then dislodged from the teeth and swallowed whole. The teeth are so long that the upper jaw has large pockets into which the teeth fit, so the fish doesn't bite itself when its mouth is closed.

Many Americans have been fooled into thinking there is a cobra on the loose by this species, the eastern hognosed snake, *Heterodon platirhinos*. When encountered, hognosed snakes will rise up, flare their ribs, and hiss, much like a cobra. Due to the hissing, many also call them "puff adders." However, hognosed snakes pose no threat to humans. In fact, if a threat persists, hognoses will flip upside down and hang their mouth open to feign death, with the hope that the potential predator will lose interest and go away.

Common throughout most of the eastern half of the United States, they occur in woodlands and fields, preferring sandy soils. Eastern hognosed snakes eat frogs, salamanders, small mammals, and invertebrates such as worms and insects. Their favorite prey is toads, and they seem to be immune to the poison produced by the parotid glands of toads. Hognoses are also known to be slightly venomous and inflict their venom using fangs positioned much further back on the jaw than most other fanged snakes. Because of this, they can often be seen gnawing on toads in order get the jaws into position to inject the venom. The mild venom relaxes the toads, causing them to deflate, making them easier for the snake to swallow.

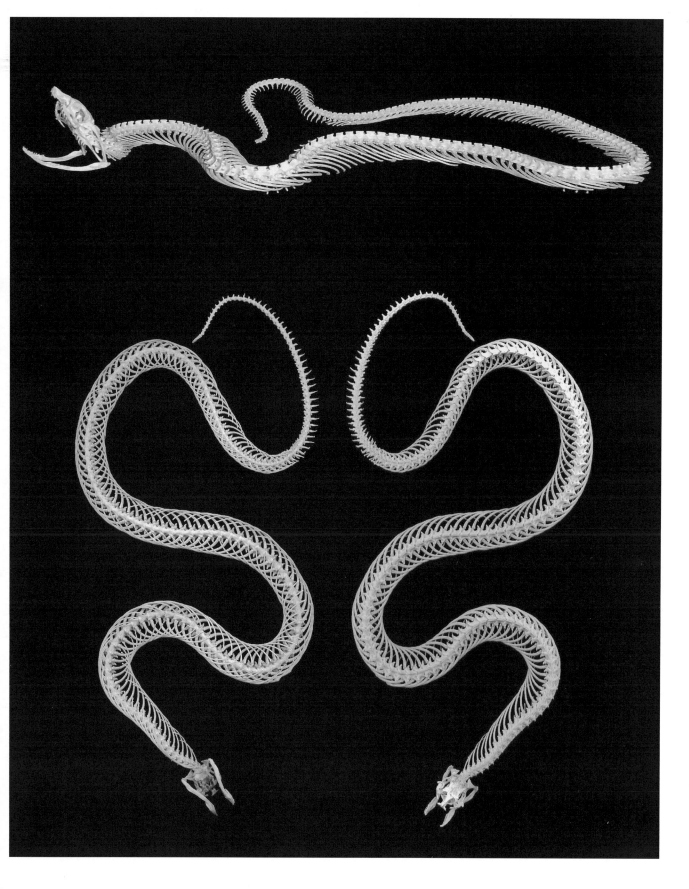

The porcupinefish, *Diodon hystrix*, which literally means "two-teeth porcupine," is a blowfish found circumtropically (around the world in tropical areas), sometimes venturing into temperate waters. They occur in sheltered caves, reefs, ledges, and wrecks. Nocturnal by nature, they tend to hide during the day and hunt at night.

Their genus name, *Diodon*, is derived from the two large dental plates that line the upper and lower jaws. They use these plates, powered by enormous muscles, to crush some of the hardest prey in the ocean, such as bivalves, snails, urchins, reef crabs, sand dollars, and hermit crabs.

Their species name, *hystrix*, comes from their ability to erect their spines, thus resembling a porcupine. When threatened, they inhale water into their stomach and inflate their body. This stretches the skin tight and causes the spines to stand up. Erect spines jab the inside of the mouth of anything making the mistake of biting a porcupinefish, and they usually get spit out to safety.

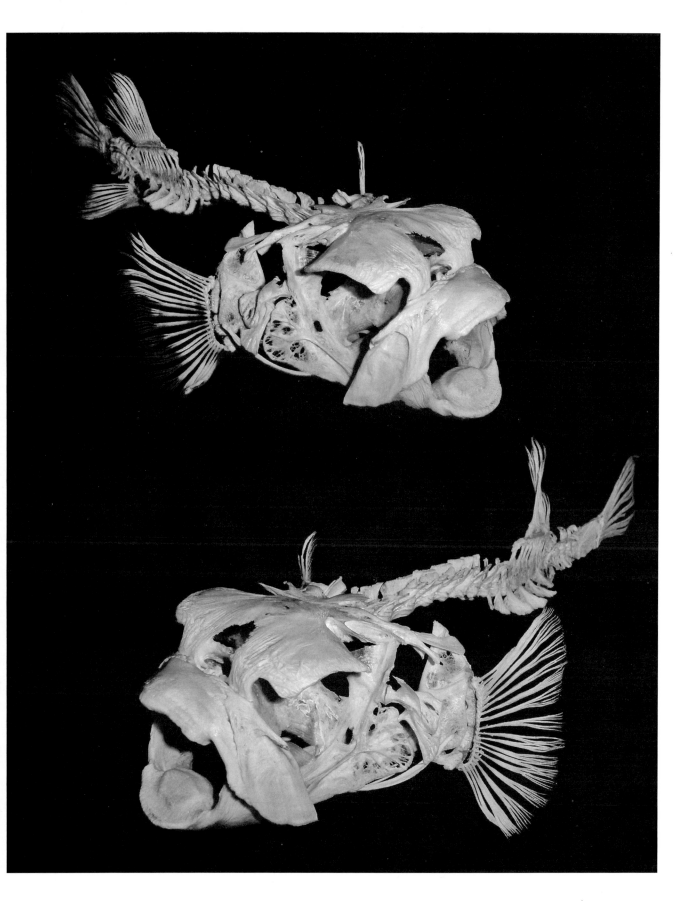

The Gila monster, *Heloderma suspectum*, is the largest lizard in North America, growing up to 22 inches long, and can be found in Arizona and Mexico. It lives in deserts and dry, rocky regions where it spends the vast majority of the time underground or at the mouth of its burrow.

Gilas are venomous lizards with toxins as strong as some rattlesnake species. However, unlike rattlers with their hypodermic-like, hollow teeth, Gilas have grooved teeth on their bottom jaw that allows venom to slowly move up the teeth and into the bite mark. Because of this, Gilas don't bite and release as most snakes do. Instead, they have very sturdy jaw muscles that help them bite, hold, and gnaw on their prey in order to give the venom time to be delivered through the wound.

Like their cousin the Mexican beaded lizard, Gilas have a suite of bony scales along their backs known as osteoderms. These serve as protection from the harsh environment in which they live, as well as from potential predators that find them too hard to chew.

■ *Melichthys vidua*, or pinktail triggerfish, is a moderately aggressive triggerfish from the Indo-Pacific, Hawaii, and East Africa. The species inhabits coral reefs, worm-rock, and rocky ledges exposed to currents in water as deep as 180 feet.

Pinktail triggerfish grow up to 15 inches long and will pair up during the breeding season to build and guard nests. They feed mainly on algae, detritus (tiny pieces of dead animals and feces), small fish, crustaceans, sponges, and octopi with their powerful, scissor-like jaws and sharp teeth.

The three spines on their back give triggerfishes their name. When a triggerfish swims into a crevice to hide from a threat, the first spine can be erected as an anchor. Next, the second spine locks underneath the first, securing it in place; this makes the first spine so stable that it will not give way unless it breaks. However, a ligament runs from the top of the third spine to the base of the second; pulling on the third spine disengages the second spine, and the first spine will drop like the hammer on a gun—this action gives the animal the name *triggerfish*.

The giant one-horned chameleon, *Trioceros melleri*, also known as the Meller's chameleon, has one large projection off its snout that points directly forward and gives this chameleon its name. It is considered a false horn, however, because it is not wrapped in a keratin sheath like the horns of many others, though sometimes a scaly point forms at its tip. This is arguably the largest species of chameleon found outside Madagascar, reaching 2 feet in length and weighing up to a pound, and is the third-largest chameleon of all behind the Oustalet's chameleon, *Furcifer oustaleti*, and the Parson's chameleon, *Calumma parsonii*.

Meller's chameleons hail from southeast Africa, including Malawi, Mozambique, and Tanzania. They inhabit tree canopies in bushy savannas up to elevations of 2,000 feet. Here, they hunt insects, smaller chameleon species, and young mammals and birds.

Like other chameleons, they are a hodge-podge of anatomical novelties. Their eyes are able to function independently, looking in two different directions simultaneously. Their feet are perfectly formed to grip branches, because they have two digits on one side and three on the other. *Trioceros melleri* have a prehensile (able to grasp) tail that acts as an additional leg. To capture food, the Meller's chameleon shoots its tongue at prey; the tip of the tongue is able to "grab" the prey, and suction is created by the middle of the tongue, holding the meal in place.

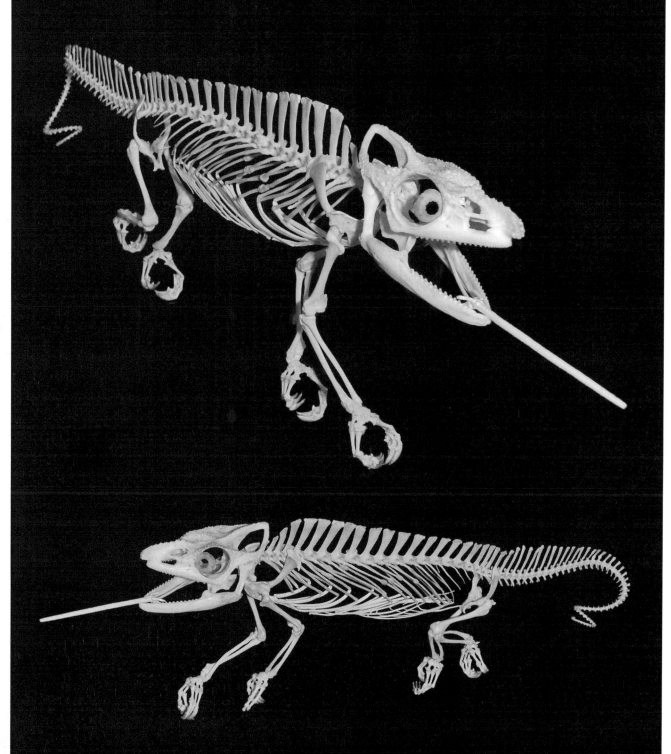

The fish world's armored car is the plated catfish, *Hoplosternum littorale*. The body is covered in overlapping plates that function as protection. Having evolved in systems with piranhas and other large Amazonian predators, they need this armor to survive the constant onslaught of toothy hunters.

The plated catfish has the widest natural distribution of all its kind, occurring in nearly all the freshwater systems of South America. It has also been introduced to southeastern Florida, where it is creating quite an ecological nightmare for endemic species because it has no natural predators and regularly raids the nests of native fishes to feed on their eggs. It also eats detritus (particles of fecal matter and dead organisms), aquatic insects, and benthic (bottom-dwelling) worms.

Plated catfish have an interesting reproductive strategy. Males build a nest made of a cluster of mucus bubbles and vegetation. Females select a nest and deposit eggs, which are fertilized by scattering sperm previously collected in her mouth. The male then very aggressively guards the eggs until they hatch.

The Hog Island boa, *Boa constrictor imperator*, was once thought to be collected to the point of complete removal of the wild populations in the Cayos Cochinos, Honduras, which is the only place they're found in the world. In the 1980s, collectors took nearly every adult from the islands for sale in the pet trade. They have since enjoyed some protections, and the population has rebounded.

This species only reaches 5 feet in length and, like all constrictors, feeds on small mammals and birds by wrapping ever-tightening coils around its prey. The human mistake of overharvesting this snake was partially alleviated by the second human mistake of introducing the brown rat, *Rattus norvegicus*, to Hog Island, which the boa pictured here has recently captured, constricted, and begun to swallow.

■ Of all the mammals, *Chinchilla chinchilla* are recognized as having one of the softest pelts. This pelt nearly brought them to extinction by human overharvesting, and their wild populations are still considered critically threatened. Their name means "little Chincha," after the Chincha people, known for wearing their pelts. Chinchillas are endemic to the Andes Mountains of Bolivia, Peru, and Chile. They inhabit rocky crevices at elevations up to 14,000 feet.

Chinchillas live in large colonies and are most active at dawn and dusk. They venture out to pursue insects and will also eat fruits, seeds, and young plant leaves. Being small and occurring in large numbers, they tend to attract the attention of many predators, such as skunks, felines, canines, snakes, and raptors. When chased, they can jump up to 6 feet in a single bound and will often spray urine to deter predators.

■

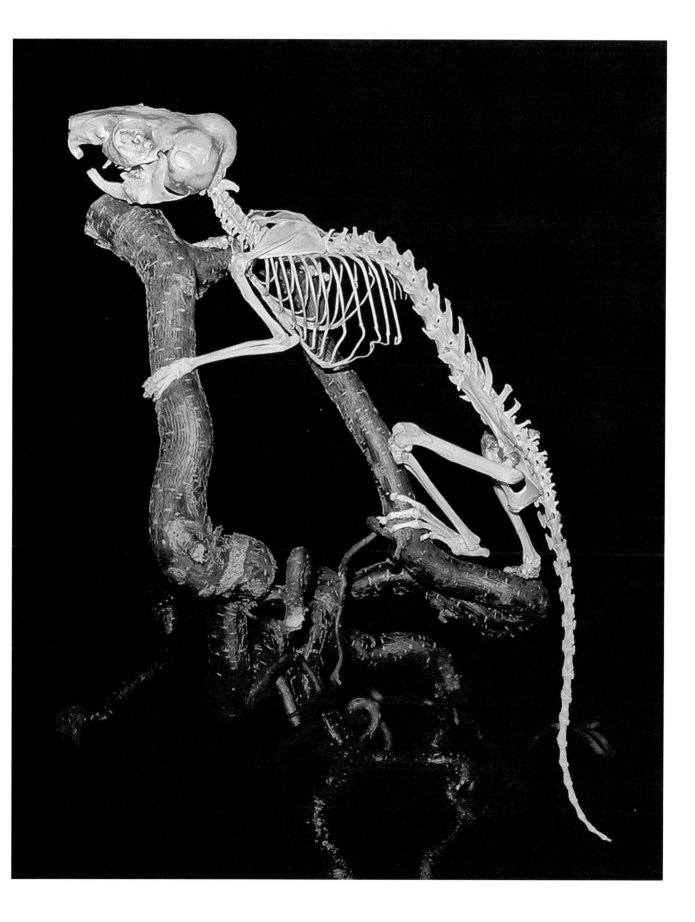

The scarlet macaw, *Ara macao*, is one of the most easily recognized of all the parrots because of its brilliant, red plumage with highlights of yellow, green, blue, and white. Common in the pet bird trade, macaws are native to a decidedly large area, from southeastern Mexico to Brazil. They live in humid rainforests, along river edges, and in savannas and are usually found alone or in pairs as they maintain monogamous pair-bonds for life. Small flocks sometimes occur, especially in places where they gather to eat clay for its minerals.

Like other parrots, macaws have very powerful beaks. They are able to process the hardest of shells and will eat fruits, nuts, and seeds. Their feet are surprisingly dexterous, as is their tongue, and they use these to handle the edible parts of whatever their beak has pulverized.

Scarlet macaws get up to 32 inches long, but most of that is their large, decorative tail. Their wing span can approach 4 feet, and they are quite capable flyers, reaching speeds up to 35 miles per hour. This speed is necessary when you could be a colorful snack for predators such as powerful harpy eagles.

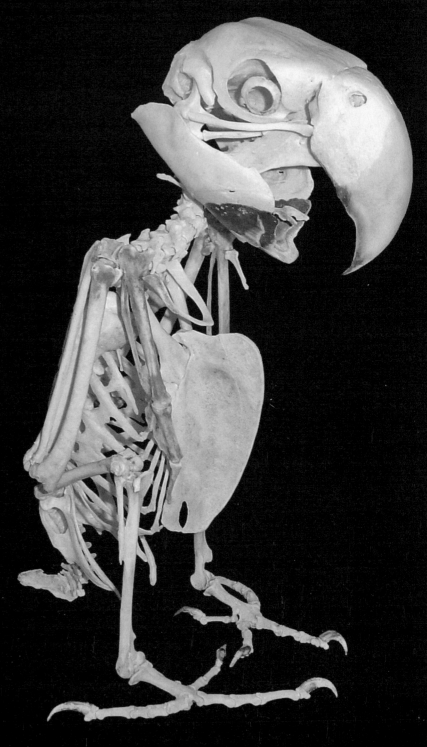

The chevron tang, *Ctenochaetus hawaiiensis*, also known as the combtooth or bristletooth tang, has a mouth full of flexible, comb-like teeth that it uses to rake detritus (small pieces of dead organisms and fecal matter) and algae. It also possesses a secondary lower jaw joint, which allows the lower jaw to open wider than in most other fishes. This expanded mouth allows chevrons to graze along the reef and sand, rasping up small bits of dead organic matter and tiny pieces of soft algae.

The chevron's large dorsal and anal fins act like a kite when the fish needs to turn sharply to avoid predators or defend home territory against rivals. Like other surgeonfishes, they have a large, scalpel-sharp spine at the base of their tail that can be erected for self-defense or fighting. The fish will stand the spine up and side-swipe rivals, with slashes that tear tissue.

The king cobra, *Ophiophagus hannah*, is the largest species of venomous snake on the planet, reaching lengths of nearly 20 feet, with reports of 25-footers encountered in the wild. It is found in India, southern China, Indonesia, and the Philippines. Kings inhabit tropical forests, plantations, grasslands, mangroves, and bamboo thickets and hunt their prey during the daytime.

Their genus name, *Ophiophagus*, means "snake eater," because kings prefer to consume other snakes. Occasionally, they will also take small mammals, lizards, frogs, and birds. They tend to envenomate their prey and begin swallowing while the prey is still alive. An 18-foot king swallowing an 8-foot dharman ratsnake, *Ptyasmucosus*, their preferred prey, is a sight to behold!

King cobras are not aggressive. Their extreme reclusiveness means that interactions with humans are much rarer than with most other cobra species. However, if cornered they will raise nearly a third of their body off the ground, flare their hood, and produce a low-frequency hiss that sounds more like a growl. A 15-foot king cobra can look a 5-foot person in the eyes when threatened. Their venom isn't as toxic as many other cobras', but the copious amount of it can be absolutely lethal. Twelve milligrams is enough to kill a human, and a king has the potential of yielding nearly 500 milligrams in a single bite.

■ Two of the most specialized feeders on the reef are the long-nosed butterflyfish, *Forcipiger longirostris*, and the forceps butterflyfish, *Forcipiger flavissimus*. The former is recognized as having the longest jaws, relative to body size, of any vertebrate. The latter is not far behind and is known to have the greatest distribution of any butterflyfish. They use these long jaws to probe the reefs of the Indo-Pacific for small invertebrates nestled in the safety of the reef structure, such as shrimps, sea stars, polychaete worms, hydroids, and amphipods.

These two butterflyfishes use their straw-shaped mouths to get close to hidden prey and then quickly crank up their buccal (mouth) pump. They do this by expanding their buccal cavity in all directions: the gills flare, the hyoid (floor of the mouth) drops, the roof of the mouth is lifted with the skull, and the tube-shaped jaws jut forward. These fast and dynamic actions all produce powerful suction that is directed out the tip of their "straw" toward their meal. This suction draws in a volume of water around their food, which in turn "drags" the prey into their mouth.

■

■ Mata mata turtles, *Chelus fimbriatus*, are underwater specialists found in the rivers, swamps, and marshes of northern South America. Their length can be nearly 20 inches and their weight up to 30 pounds. They inhabit mainly shallow waters where, because of their long neck, they can reach the surface for a breath without swimming up from the bottom and giving away their position to predators such as caiman.

Mata matas are perfectly designed for river life and for ambush hunting. Their shell is flattened, which allows water to flow over it more easily. Both their shells and skin are patterned in diamond and triangle shapes that blend perfectly with the fallen leaves in which they hide.

Matas sit motionless with their frilled neck slightly extended into the water. When unwary prey get too close, they explode their mouth open and expand the hyoid bones of the throat. This generates a violent inward flow of water, carrying the prey into their mouth. The jaws are then slightly opened, allowing the water to gush out and the prey, including mainly fish but also amphibians, insects, and crustaceans, to be swallowed whole.

■

The aptly named stonefish, *Synanceia verrucosa*, is considered the world's most venomous fish and perfectly resembles a rock lying on the ocean floor. They inhabit reefs, lagoons, and rocky areas in water up to 100 feet deep throughout the Indo-Pacific. This species can grow up to 16 inches long and has 13 stout, grooved spines in their dorsal fin. Each spine has a venom sac at its base; the venom travels up the grooves when the spines penetrate something, flesh or otherwise. This system evolved to deter predators, but sometimes it is employed when a diver or wader accidentally steps on a stonefish. Envenomations by stonefish are rarely fatal, though they can be excruciatingly painful.

Stonefish are amazingly cryptic, with skin colors that resemble coralline algae, sponges, and rocks. They sometimes grow live algae on their skin as well. Their skin also has many fleshy lumps and protuberances to break up their outline. To ambush prey, they either lie in wait or "waddle" along the reef structure, swaying with the waves to creep closer to a meal. Then, they launch one of the fastest strikes of all vertebrates—from mouth closed to fully open to closed again in as little as 14 milliseconds!

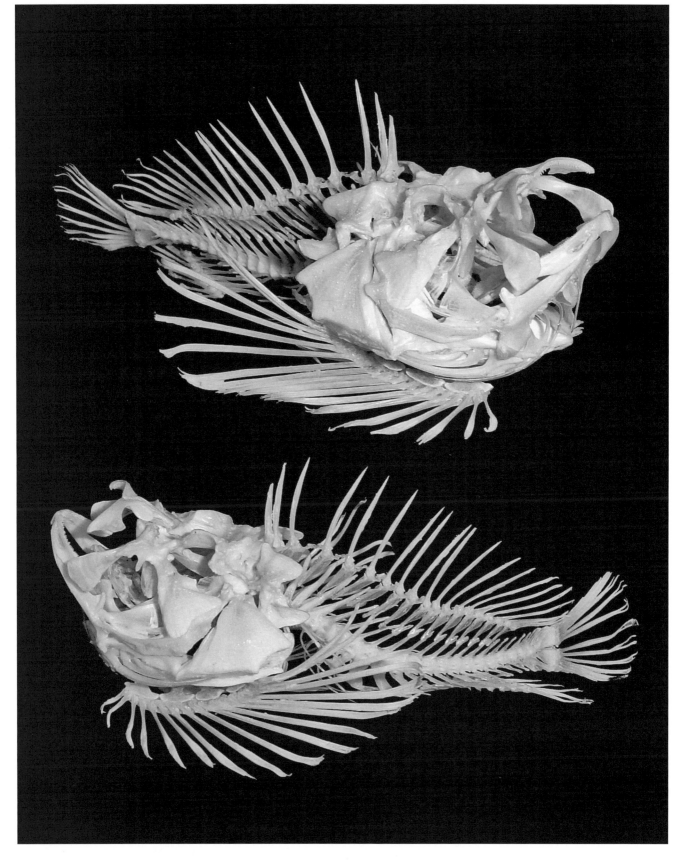

The Ethiopian mountain adder, *Bitis parviocula*, was only discovered in 1976 and has the narrowest range of all the *Bitis* genera (plural of "genus"), which includes puff adders, rhino vipers, and gaboon vipers. It is a fat-bodied snake, rarely reaching 4 feet long. This species is found in Ethiopia at elevations from 5,000 to nearly 10,000 feet. The habitat preferences appear to be extremely broad, from montane (moist, cool, sloping, and below timberline) to grassland to cleared plantation. However, only a few specimens have been found, so our knowledge is limited.

Ethiopian mountain adders are believed to be nocturnal and, like their cousins, probably eat small mammals that they capture using an ambush technique. Their brilliant green, brown, yellow, and black skin, littered with triangles, squares, and hexagons, makes hiding very easy.

As a *Bitis* species, they have extremely long fangs capable of delivering copious amounts of venom. What little is known about the venom suggests it can be lethal to humans, especially since most bites occur at elevations far from medical facilities. Like other venomous animals, they are completely harmless if left alone and serve a critical ecological function in the wild as population control for rodents.

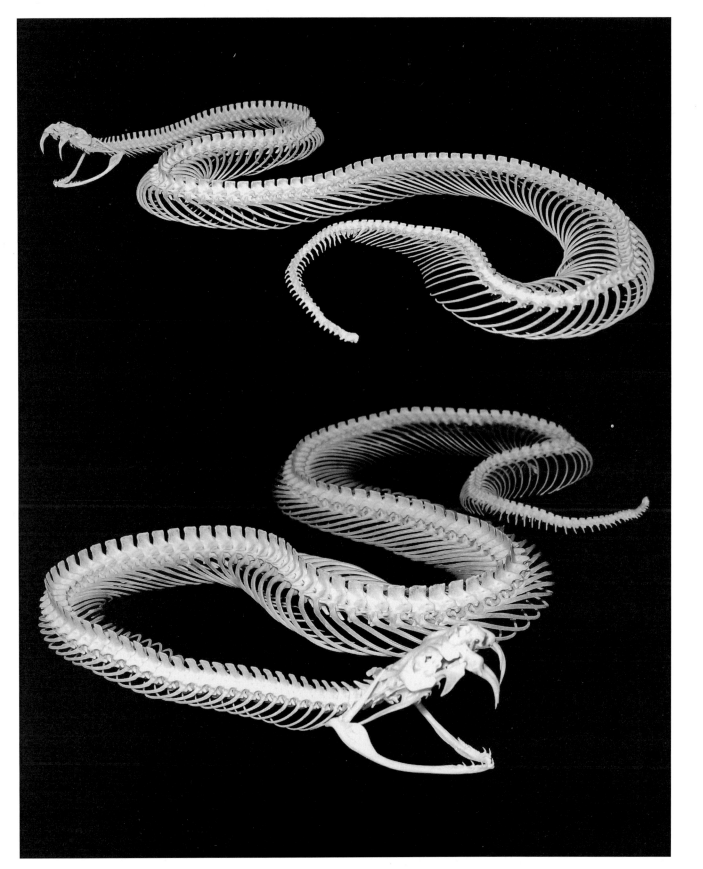

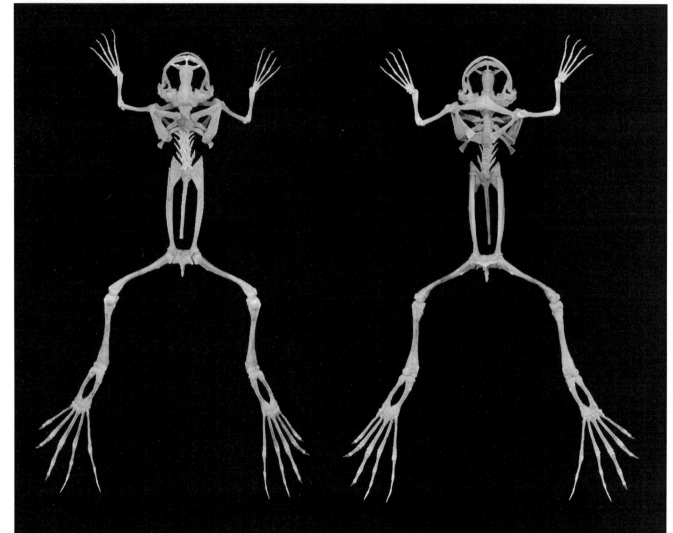

The African clawed frog, *Xenopus laevis*, is native to eastern and southern Africa near the Rift Valley. It lives in stagnant pools and quiet streams, surfacing only to breathe and when forced to migrate if their habitat dries out for too long. Three of their rear toes have a horny, claw-like tip, and hence the name "clawed frog," but they're not true claws. Their front feet have no webbing, allowing them to use their fingers to cram food into their insatiable mouths. They have a well-developed lateral line system that is highly sensitive to water movement, helping them find live prey. Their diet includes anything dead or alive, such as insects, worms, crustaceans, fish, tadpoles, and snails.

Since they live in a region prone to drying out for extended periods of time, African clawed frogs can burrow into the mud and lie dormant for up to one year. If rain does not return after a year, they will climb out and crawl (they can't hop) to another water source. They have been introduced all over the world and in many places are harming the delicate natural ecosystem. Interestingly, this was the first vertebrate species to be completely cloned in the laboratory, and they are used extensively in embryology research.

The largest wrasse species found in the western Atlantic is the hogfish, *Lachnolaimus maximus*. They grow up to 3 feet long, weigh up to 24 pounds, and are easily recognized by the 3 or 4 dorsal filaments that stand high off their back when extended.

Hogfish inhabit worm-rock structures, coastal reefs, open bottoms, and rubble piles in water up to 100 feet deep. This species is prized for its subtle, white, flaky meat and is targeted by anglers and spear fishers alike. Their flesh is often used to make ceviche, a popular raw appetizer.

Hogfish are considered specialized molluscivores, which means they eat mainly clams, but they'll also take snails, crabs, urchins, shrimps, and small fish. Their oral jaws are capable of opening very wide and are tipped with long teeth shaped like canines, which they use to "rake" the ocean floor for buried bivalves (molluscs with a hinged shell). Food is then scooped up and moved to the throat, where they have powerful pharyngeal (meaning "in the pharynx," the area between the mouth and the esophagus) jaws covered with thick, heavy teeth for crushing hard prey.

■ Leopard geckos, *Eublepharis macularius*, are relatively large geckos (up to 11 inches long) endemic to the deserts of Pakistan, Afghanistan, Iran, and northern India. They get their name from their spotted, leopard-like appearance. Leopard geckos prefer habitats such as rocky, dry grasslands, arid scrublands, and deserts where they spend most of the daylight hours underground, avoiding the heat and predators. They are most active during dawn and dusk, when the weather is conducive to hunting.

Leopard geckos eat insects of all types, which they capture through quick pursuits and snappy jaws lined with many teeth. They cannot climb smooth surfaces like many other gecko species, so they must hunt mainly on the ground. However, unlike other gecko species, leopard geckos have eyelids, which they can close for protection when sand is blowing.

Snakes, raptors, predatory mammals, and other reptiles prey on leopard geckos. Luckily, these geckos possess the ability to "drop" their tails when threatened by a predator. The tail pops off and may continue to wiggle on the ground for many minutes after falling off. This gives the predator something to keep busy with, while the gecko scurries away to safety. The gecko will then regrow the tail after some time, only without vertebrae inside. Rather, a sheath of cartilage will form and grow, with muscles and fat deposited around it.

■

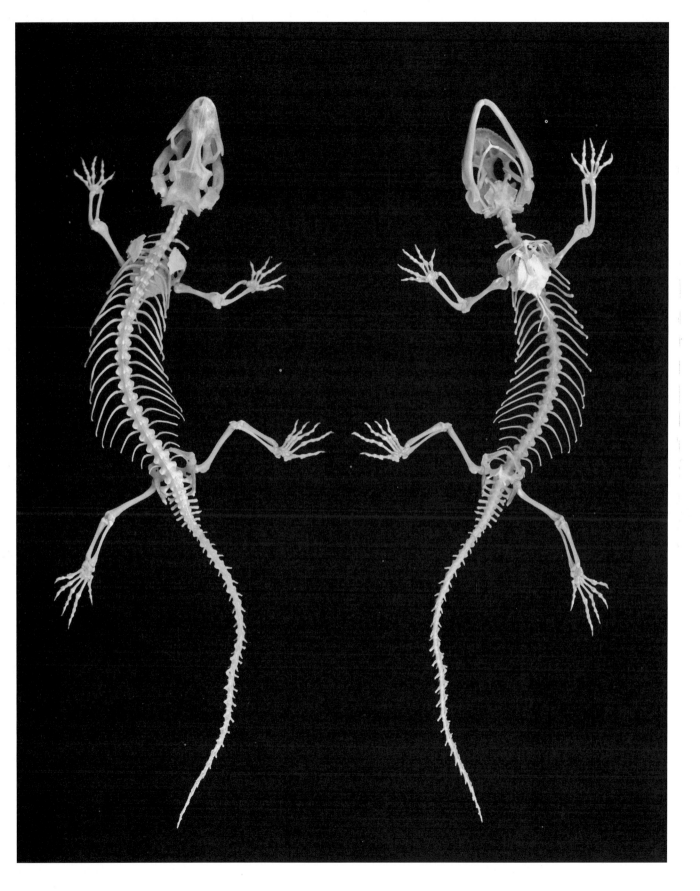

The silver arowana, *Osteoglossum bicirrhosum*, is a freshwater predator found in the Amazon basin. It occurs in rivers and streams with relatively clear water. The genus name, *Osteoglossum*, means "bony tongue," and the species name, *bicirrhosum*, means "two barbels," referring to the two fleshy projections on the lower jaw.

Silver arowana are famous for their acrobatic feeding style. They are happy to slurp up frogs, fish, and insects that reside on the water's surface but are also willing to leap up and grab birds, bats, and lizards off overlying branches. Their powerful bodies, with extremely long anal and dorsal fins, propel them effortlessly through water or air. Their large mouths, lined with dozens of cone-shaped teeth, easily grab prey as big as their head. Capturing a meal by leaping into the air has earned them the common name *monkeyfish* for many local tribes.

At up to 26 inches long and 14 pounds, *Canthidermis sufflamen*, or the ocean triggerfish, is the largest triggerfish in the western Atlantic. The species inhabits outer reef slopes and rocky ledges that border deep, open ocean. They are perfectly comfortable in water over 1,000 feet deep, where they hang out near floating mats of *Sargassum* seaweed or other ocean-borne debris. They will even reside along an anchored fishing boat if the boat stays in one place long enough.

Ocean triggerfish have large, perfectly opposable dorsal and anal fins that they use to propel themselves through the water. A "wave" is propagated down the length of the fin from front to rear, and this motion serves to push the fish forward by generating thrust. These fins also make ocean triggers highly maneuverable, because they can reverse the "wave" and swim backward, something very few other fishes can do. Ocean triggerfish need this maneuverability to hover in front of the zooplankton and small, neutrally buoyant invertebrates on which they feed.

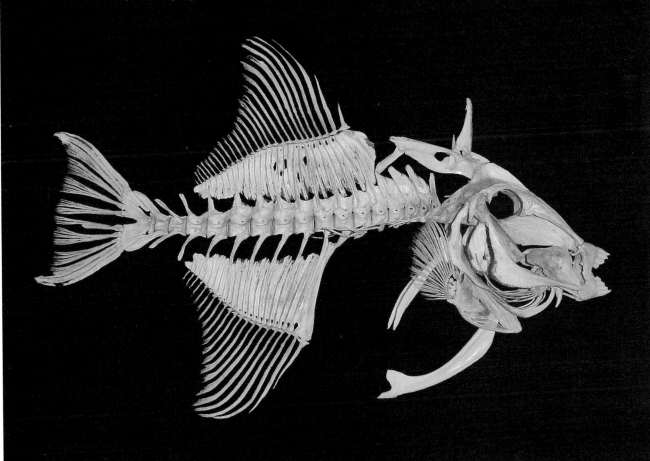

The longlure frogfish, *Antennarius multiocellatus*, is a reef-dwelling fish found in the western Atlantic from Bermuda to Brazil. The giant frogfish, *Antennarius commersoni*, is found in the Indian and Pacific Oceans. Both inhabit warm, shallow reefs with abundant marine sponges. Their skin is colored and textured like that of reef sponges so they can be perfectly concealed as they hunt for food.

Frogfish have short, fat bodies, giant mouths, and odd pelvic and pectoral fins. Their fins function like hands or arms, allowing the fish to "crawl" along the reef or position themselves in reef crevices, sometimes upside down. This helps them get into position to catch their prey, which can be larger than themselves; they have highly mobile jaw and skull bones that allow them to capture and swallow huge prey. This is a necessity when you may only get to feed once a week or even less often.

The frogfish is a fish that fishes for other fishes. It does so with a highly specialized, modified, dorsal spine called the illicium. At its tip is a fleshy appendage called the esca that wiggles with the action of the illicium. The frogfish can use the esca to attract the attention of unsuspecting small fish or shrimp, thinking it might be a nice meal of plankton. When they get close, the frogfish unleashes one of the fastest feeding strikes on the planet—the entire feeding cycle can take less than 20 milliseconds. The average human eye blink takes longer than 100 milliseconds, so this fish could feed at least five times in the time it takes you to blink your eye.

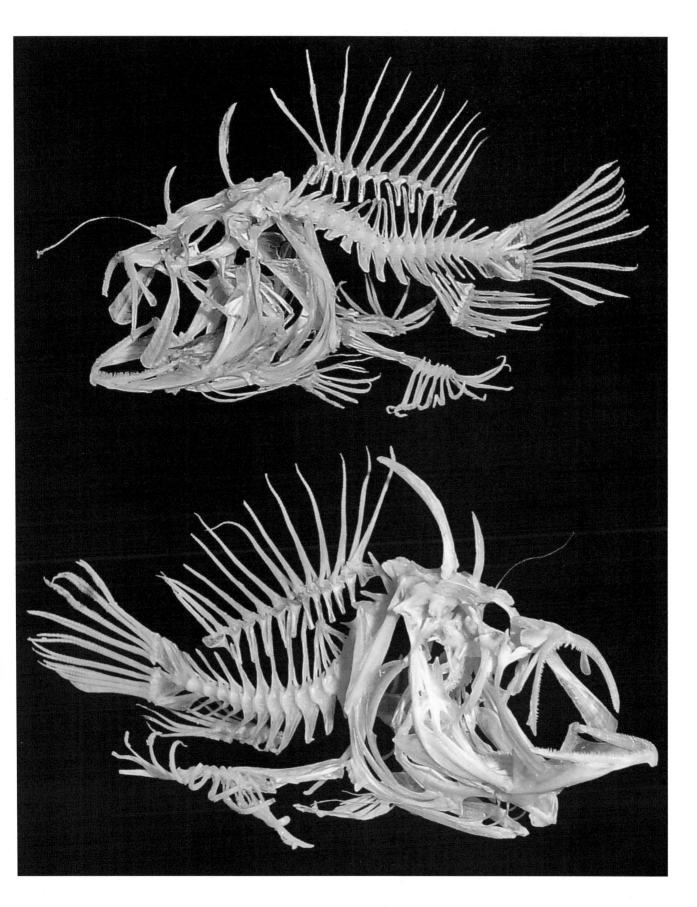

The eastern kingsnake, *Lampropeltis getula*, is found throughout the eastern United States as far north as New Jersey. It inhabits rocky areas, brushy hillsides, river valleys, woodlands, open fields, swamps, and some forests. Kingsnakes are diurnal and are often encountered by humans, to whom they pose no risk. In fact, many species are kept as pets because of their willingness to be handled and their nonaggressive nature.

Eastern kingsnakes grow up to 5 feet long and commonly eat small mammals, birds, lizards, frogs, and turtle eggs. However, they are best known for their ability to consume pit vipers—rattlesnakes, such as this massasauga rattlesnake, *Sistrurus catenatus*, copperheads, and cottonmouths. Kings, as they are called because of this ability to eat venomous snakes, are entirely immune to the venom of these other snakes. They track down another snake and latch onto it with a bite from their mouth, full of four rows of teeth on top and two rows on the bottom; then they constrict around it to kill and swallow it head first when it stops struggling.

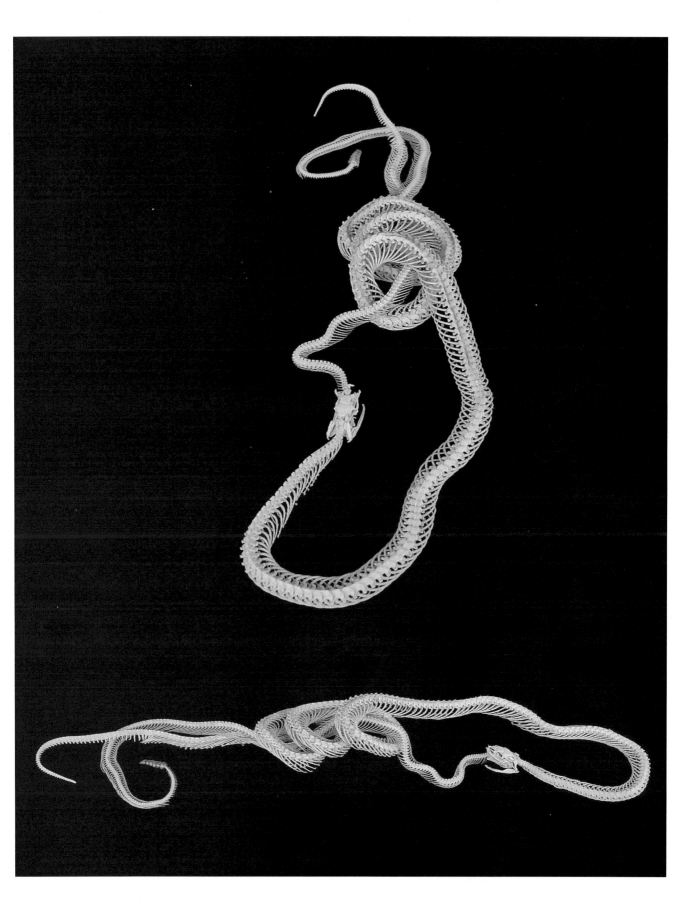

America's most common catfish species, the channel catfish *Ictalurus punctatus* is perfectly built for life as an omnivore in search of anything living or dead. The tubular body shape and forked tail helps it swim through its preferred waterways—rivers and streams—though it can also be found in lakes, reservoirs, and ponds.

Like all catfishes, they lack scales. This allows their body to be covered in taste buds, so the entire skin surface acts like a giant tongue, able to taste the water for food. This is also why smelly bait can be so successful at attracting catfish. Their large mouths make it possible to engulf prey nearly as wide as their head, such as this rockbass, *Ambloplites*

rupestris, and this bluegill sunfish, *Lepomis macrochirus*.

Channel cats, like other catfishes, have barbels, or "whiskers," around their mouth that are used to taste the area near the mouth when locating prey. These whiskers cannot sting, contrary to popular belief. The real threat comes from the barbed spines on the dorsal and pectoral fins.

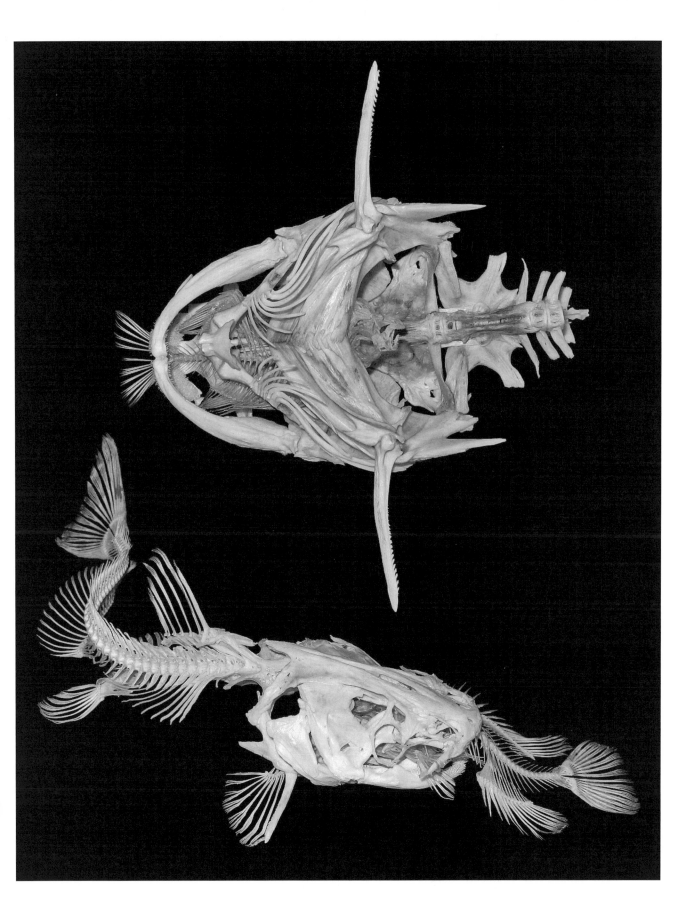

The rhinoceros viper, *Bitis nasicornis*, is a fat-bodied snake found in the tropical rainforests of Central and Western Africa. It inhabits lowland, swampy environments with water nearby. The common name comes from the scaly projections that rise above each of the nostrils, creating the effect of small horns.

Rhinoceros vipers grow up to 4 feet long and can be 6 inches wide at their thickest point. They have one of the most remarkable skin patterns of all snakes. Cryptic (camouflaging) coloration helps them hide from prey while they lie in wait to ambush small mammals, lizards, amphibians, and occasionally fish. Their venom attacks the circulatory system, causing blood cells and vessels to hemorrhage.

Like other members of the viper family, their fangs are mobile, meaning they can swing out on hinges to pierce their prey or any predator that threatens them. They are not considered aggressive snakes and usually provide plenty of warning when they feel threatened by puffing up (by inhaling extra air and spreading their ribs) and hissing loudly.

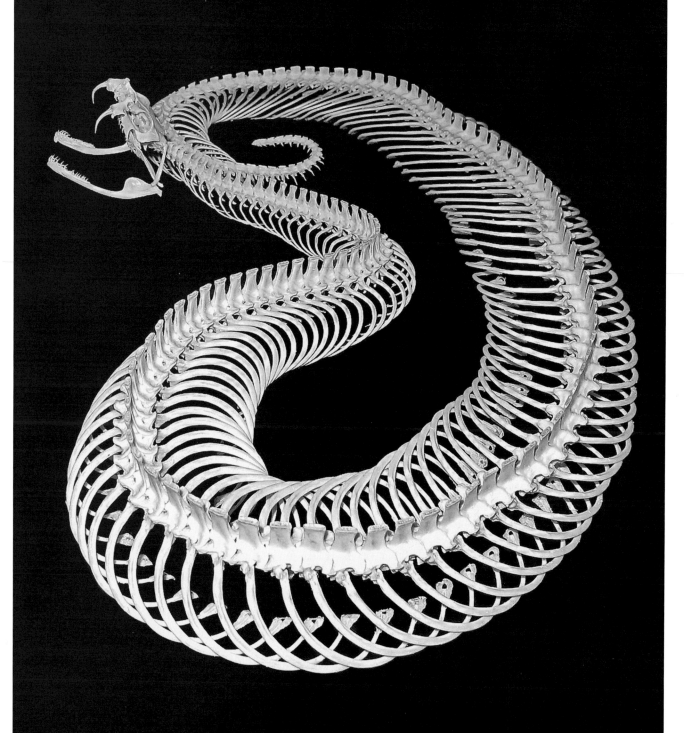

A large grouper species from the Indo-Pacific, *Epinephelus tukula* is also known as the potato grouper or potato cod. They can grow up to 5 feet long and weigh as much as 200 pounds. Though the species is widely distributed, their populations tend to be patchy due to their propensity to gather into breeding aggregations. They are also known to exhibit extreme territoriality and site fidelity, which makes them easy to catch or spear and prone to overfishing.

Potato groupers are one of the top predators on the reefs where they reside. They are usually found in reef channels and near seamounts (mountains rising from the ocean floor but not reaching the surface) where water currents move quickly. Here, they establish territories and defend them aggressively against intruders. It's also from here that they launch attacks on unwary fish, crustaceans, and even baby sea turtles. Their pointed heads make for easy swimming in the current, and their large mouths can engulf prey almost as big as their own head.

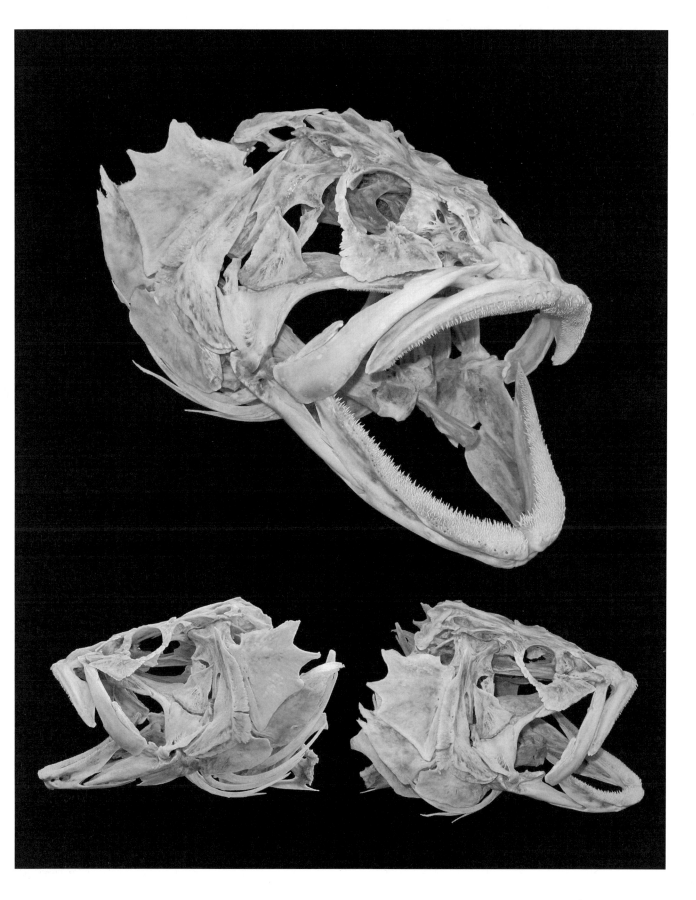

■ The copperband butterflyfish, *Chelmon rostratus*, and the beaked coralfish, *Chelmon marginalis*, are just two of the highly specialized species within the butterflyfish family, Chaetodontidae. They live next to coral reefs and rocky shores of the Western Pacific and sometimes venture into estuaries in search of habitat or food. The highly specialized jaws of *Chelmon* make them perfectly suited to probe the reef for small invertebrates, such as crustacean amphipods and isopods, shrimps, and worms. Their jaws function much like a straw, creating a small opening at the end of a tube. As the mouth generates suction, the jaws point it where it's needed to suck in their meal.

Found in the coastal regions of southern Africa, the green mamba, *Dendroaspis angusticeps*, is a fully arboreal (tree-living) snake species. It appears to prefer tropical rainforests in coastal lowlands, but some have been found in coastal dunes and montane forests (moist, cool slopes below the timberline). Their long, slender bodies are perfect for sliding through branches, and special processes on their vertebrae let some of the muscles create cantilevers, allowing the snake to bridge large gaps in the canopy.

Green mambas spend their entire lives in the tree canopy, hiding among the green leaves, into which they blend perfectly. They venture to the ground only in pursuit of obvious prey. However, unlike most other snakes from the elapid family, which actively hunt for their food, green mambas appear to be willing to sit and wait for prey to come to them. Their typical meal includes bats, birds, eggs, and the occasional small mammal.

Though mambas are members of the family known to have fixed fangs, which means they do not swing on "hinges" like those of snakes such as vipers and rattlers, green mambas have slightly mobile fangs. Their fangs do not fold back when not in use, but they can be projected outward to assist in envenomating their prey. The venom delivers a mix of neurotoxic (toxic to nerve cells) chemicals and a compound known as dendrotoxin. Dendro-toxin is the most fast-acting snake venom toxin known and, though the species is not perceived as a threat due to its reclusive nature, bites are often fatal.

The eastern smooth boxfish, *Anoplocapros inermis,* occurs in waters up to 1,000 feet deep along reefs and harbors of the southeastern Australian coast and elsewhere in the Indo-West Pacific. It can grow to 14 inches long and is a slow, cumbersome fish that seem to scuttle along the bottom in search of benthic (living at the bottom of the sea) prey, such as algae, shrimps, worms, bivalves, snails, and tunicates.

As members of the fish family Ostraciidae, they have an interesting hard body covering called a cuirass, composed of fused, triangular, dermal plates. This structure is like a suit of armor and provides protection from predators. However, it also compromises their ability to flex their bodies, which slows them down during swimming. Therefore, they swim by flapping their pectoral fins, employing their tails only during faster swimming to escape threats.

This family is known to excrete a toxin called ostracitoxin in their mucus when threatened. The toxin is hemolytic, meaning that it destroys red blood cells and releases hemoglobin, thereby eliminating the blood's ability to transport oxygen to the tissues.

A brilliantly colored species of fish from the western Pacific Ocean, Harlequin tuskfish, or *Choerodon fasciatus*, are characterized by bars of blue and orange along their body. Their common name arises from their large, blue, tusk-like teeth. They usually live in water less than 100 feet deep and can grow up to 12 inches long.

As a member of the wrasse family, Labridae, they swim mainly by pectoral-fin flapping, wherein they beat their side fins and seem to bounce along the water just above a reef. Something else this family is known for is a sophisticated second set of jaws, called pharyngeal jaws, in their throats. These are usually lined with teeth that perfectly match the fish's chewing needs. In the case of Harlequins, which like to eat urchins, sea stars, clams, snails, and small crabs, these teeth are shaped like molars and are used for crushing hard food before it is swallowed.

■ The largemouth bass, *Micropterus salmoides*, is the most sought-after sport fish in the world. The amount of money spent on bass boats, fuel, fishing licenses, tournament fees, tackle, and gear by bass anglers totals more than a billion dollars annually, and fishing for this species is now a global endeavor.

Largemouths are endemic to the eastern half of the United States and occur in every conceivable freshwater habitat, from bayous to creeks and from lakes to reservoirs. The world record is a largemouth weighing 22 pounds 4 ounces, and the potential windfall for the person who catches the next world record is immeasurable.

Largemouth bass are the white rat of fish-feeding studies. Their voracious appetite and willingness to feed after being surgically manipulated has led to significant insights into the evolution of aquatic feeding. They are excellent ram-feeders, which means they overtake prey by out-swimming it and engulfing it with their huge mouth. Remarkably, they can also feed using suction because (1) the sides of their mouth are blocked by bones and soft tissue, which makes for a large, round opening perfect for sucking in their prey, and (2) they possess the ability to modulate their behavior from mostly ram-feeding to mostly suction-feeding, depending on the prey they're after and the environment in which they're hunting.

The copperhead, *Agkistrodon contortrix*, is one of North America's most commonly encountered venomous snakes. Like their close relative the cottonmouth, they tend to be rather defensive snakes prone to striking when encountered. However, copperheads have the least toxic venom of all the American pit vipers, and human deaths from their bite are very rare.

Copperheads are found in the eastern half of the United States in wooded habitats where their camouflage blends in perfectly with leaf litter. They are also found on rocky hillsides, where they often bask on the open rocks. When feeling threatened, a copperhead will gather into a tight coil and twitch its tail vigor-ously in the dried leaves, generating a sound resembling the rattle of a rattlesnake. They usually give passersby plenty of warning and most often flee long before you may know they were there.

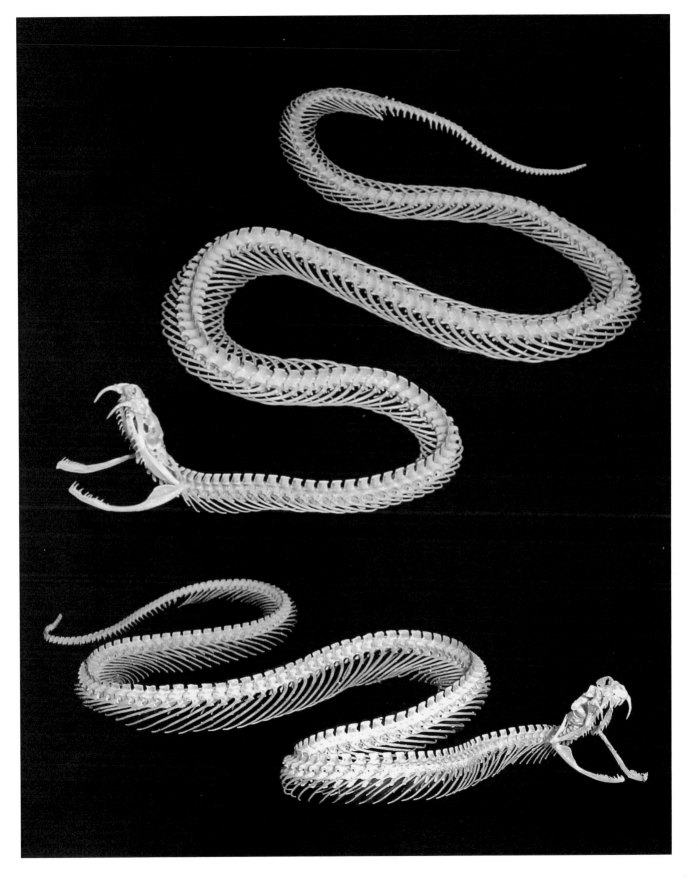

■ Moorish idols, *Zanclus cornutus*, are tropical, marine fish common throughout the Indo- and eastern Pacific. Their habitat is lagoons, coral reefs, and rocky outcroppings in water up to 500 feet deep. They commonly occur in small, tightly grouped schools of two to five individuals, where their zebra stripes help them blend together, making it difficult for a predator to determine where one idol ends and another starts. An idol's long, dorsal fin is thought to trick predators by giving these small fish a much larger body profile, making bigger fish think they're too large to swallow.

Moorish idols are close relatives of surgeonfishes and tangs, but they lack the scalpel-like spines on the tails of their relatives. They are the only members of their family, Zanclidae. The name is derived from the Greek word *zanclon* and translates into "sickle shape."

Omnivores, their long jaws and pointed face allow them to probe reef crevices for small, hidden invertebrates, while their small, comb-like teeth make it possible for them to eat soft food such as sponges, algae, and the particles of dead organisms and feces known as detritus.

■

■ The tropical rainforests of northern South America are the home of prehensile-tailed porcupines, *Coendou prehensilis*. All but their tail is covered in short, thick quills for protection from predators such as jaguars. When threatened, they sit back, stomp their feet, shake their coat, and growl and squeak at their pursuer. They will readily sink their long teeth into any captor if curling into a ball for protection doesn't work first.

Prehensile-tailed porcupines are herbivores and spend nearly all their time in trees, where they eat leaves, buds, flowers, fruits, some nuts, and the cork found under tree bark. Their huge incisor teeth function like scissor blades, clipping vegetation that will then be ground by their large, flat molars in the back of their jaws. Adept climbers, their feet are modified for grasping and their tail functions like a fifth limb, capable of gripping a branch firmly enough that they can hang by it while they reach for food.

■

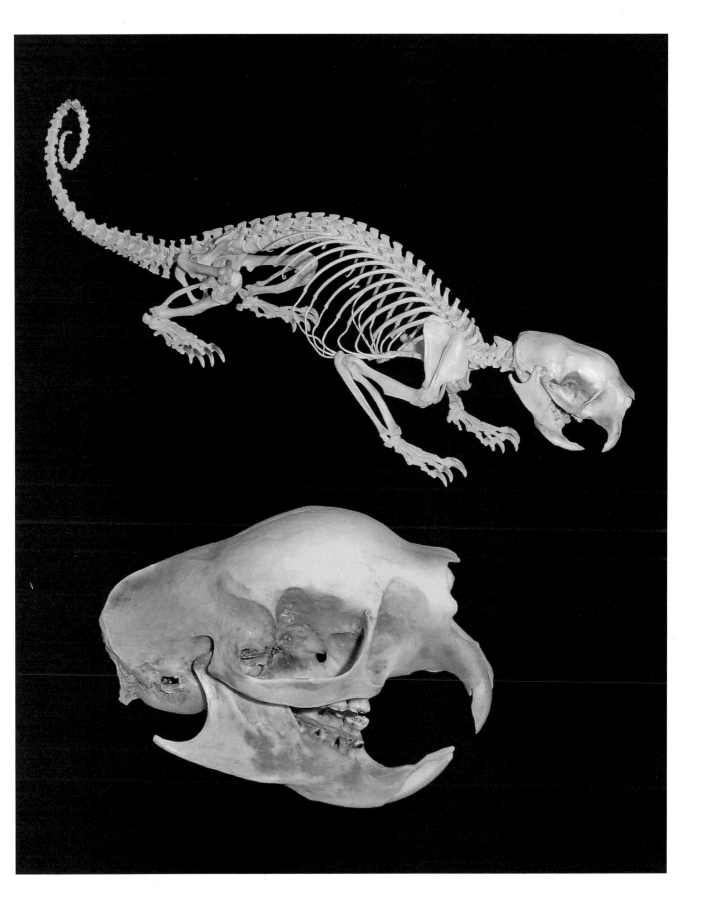

■ Bell's hinge-back tortoise, *Kinixys belliana*, is a turtle species found in the tropical and sub-tropical savannas of sub-Saharan Africa. It frequents forest edges next to water, where it grabs millipedes, snails, slugs, worms, tadpoles, and lush vegetation for food.

As their name implies, all hinge-back tortoises have a hinge in the carapace, or shell, that allows the back portion of it to close over the rear legs and tail for protection. The head and forelimbs have much thicker scales than the back legs and tail and can be tucked inside the shell for protection as well. When threatened, they will often huddle the front of their body against a tree trunk or root and close the back part of the shell in defense.

This species, and other hinge-backs, are severely threatened in their native environments, due to habitat fragmentation and over-harvesting for bush meat.

■ Male Jackson's chameleons, *Trioceros jacksonii*, have three *Triceratops*-like horns on their head that point directly forward and may curve slightly. Females usually lack horns, though variation in female head adornment has been observed, from slightly raised scales to full horns. The species gets up to 14 inches long and is highly territorial, with males using their horns to shove each other around in displays of dominance.

Jackson's chameleons hail from Kenya and northern Tanzania and have been introduced to Hawaii in recent years, where they are thriving. They inhabit cool, humid slopes below the timberline in forests up to elevations of 8,500 feet. As such, they are capable of dealing with weather conditions that vary, from nearly 40°F to 86°F where humidity levels are as high as 80%.

Like other chameleons, they exhibit many anatomical novelties. Their eyes can work independently, looking in two different directions at the same time. Their feet have two digits on one side and three on the other, an adaptation that lets them grasp branches perfectly. Their prehensile tails function as a fifth leg. Their tongues are missiles that get shot at their food and stick by "grabbing" prey with the tip of the tongue; the middle of the tongue creates suction to hold their meal, made up primarily of snails and insects.

The redtail boa, *Boa constrictor constrictor*, uses its forked tongue to track the smells of prey. The tongue's fork creates a left and right side, allowing the snake to determine which direction has the greatest concentration of odors. After flicking their tongue in the air and sampling odorants, the tongue is brought back into the mouth and the forks are tucked into the two-sided vomeronasal organ (a.k.a. Jacobson's organ), which determines both whether the smell is from something they can eat and which direction the snake needs to go to find it.

Boas capture their prey via fast strikes, where the prey is snatched up between the jaws and immediately wrapped in coils that stop blood flow, like the one placed around this grey squirrel, *Sciurus carolinensis*. Once dead, prey are swallowed head first and whole.

Contrary to popular belief, snake jaws do not "dislocate" or "unhinge." Rather, (1) the lower jaw has no symphysis (joint) between the left and right sides; (2) the lower jaw joint has a large range of motion; and (3) the bones (quadrate and supratemporal) that attach the lower jaw to the skull have loose connections to the back of the head. These three features allow the jaws to open wider than the head, so the snake can eat prey of greater girth than itself.

Bothus lunatus, or the peacock flounder, is a flatfish found in the coastal, tropical, and subtropical waters of the eastern and western Atlantic Ocean. It occurs in water less than 300 feet deep and is most common in water less than 75 feet deep. This species is the most common flounder encountered on reefs, but it also inhabits mangroves, seagrass beds, and rocky piles. The common name comes from the light-blue spots located on the pigmented top side.

Peacock flounder begin life like any other larval fish—bilaterally symmetrical, with one eye on each side of the head. However, very early in their lives the right eye migrates over the top of the skull, and thus two eyes exist on the left side. At the same time, the fish turns on its side to create an eyed side that watches the water above and an eyeless side that lies on the ocean floor.

Peacock flounder have small gut cavities and completely lack a swim bladder. This makes swimming along the bottom easier, and here they pursue their fish, crustacean, and octopi prey. This species can grow to 18 inches in length and is a prized sport fish for the enormous amount of meat they yield, relative to their body size, when filleted.

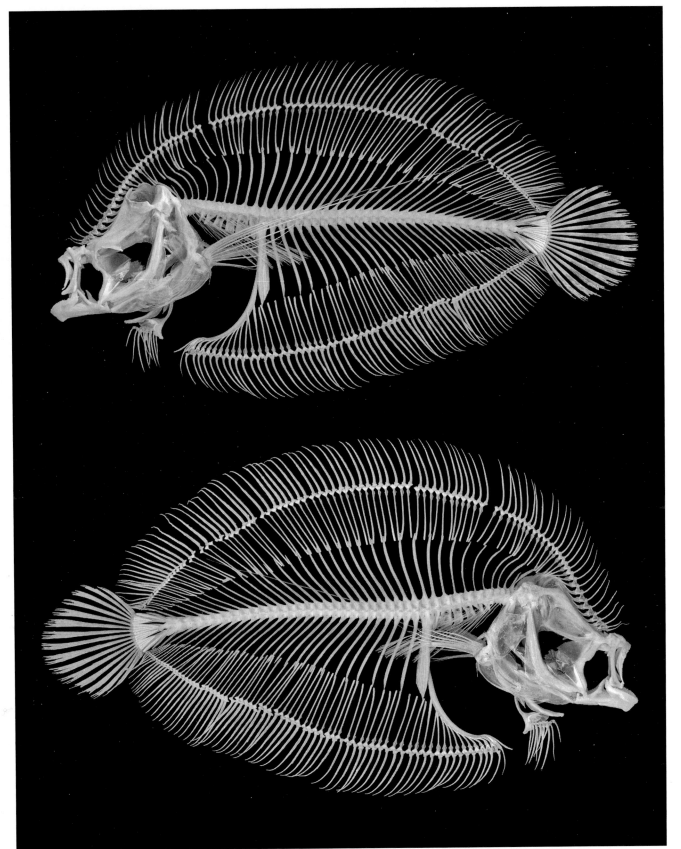

■ One of the ocean's most intriguing fishes, the lined seahorse, *Hippocampus erectus*, is adored by all. It inhabits mangroves, seagrasses, sponges, corals, and floating *Sargassum* of the western Atlantic, from Nova Scotia to Uruguay.

Seahorses form monogamous pairs, and the female eventually deposits her fertilized eggs into a pouch at the base of the male's abdomen. Here they develop until he ejects half-inch-long baby seahorses into the water column to fend for themselves.

Seahorses have prehensile tails that they use to hold on to structures. Being poor swimmers, this ability is highly useful in currents or habitats with wave action. They also use their prehensile tail to creep up on and position themselves in front of planktonic prey such as amphipods, copepods, decapod larvae, and very small shrimps. To eat, they snap their head up extremely fast, which serves to expand their mouth cavity three dimensionally, creating the perfect suction pump to slurp prey.

Scaleless, seahorses are made of an interlocking latticework of bony rings. Unfortunately, their dried bodies have become one focus of a culture of medicinal myth, where some believe in the healing or libido-enhancing powers of powdered seahorses. There is absolutely no scientific evidence to support this, and yet it has led to the demise of many seahorse populations.

■

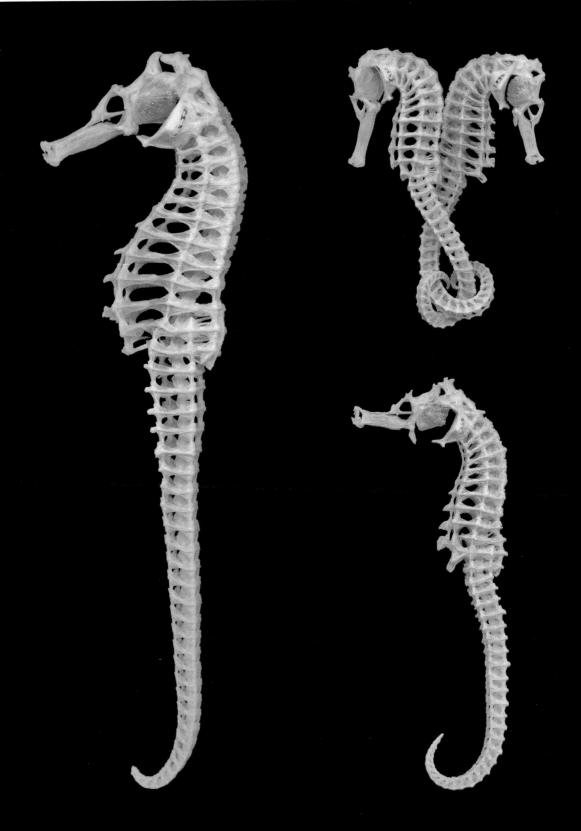

■ The sharp-nosed pit viper, *Deinagkistrodon acutus*, is a highly venomous, 5-foot-long snake from southern China, Taiwan, Viet Nam, and Laos. It is found on mountain slopes and rocky hills over 4,000 feet high. The species has another common name, the hundred-pace pit viper, because a bite victim supposedly has only 100 paces left before they die. Of course, this is an exaggeration. That said, this species can be fatal with its highly hemotoxic venom, which leads to almost immediate hemorrhaging, swelling, pain, and blistering.

Sharp-nosed pit vipers are nocturnal (active at night) and hunt warm-blooded rodents and birds using heat-sensing pits on their snout. In contrast, they hunt cold-blooded frogs and lizards by their smell and movement. The species is known to have relatively long fangs that deliver large quantities of venom, and they are ill-tempered enough to use them.

■

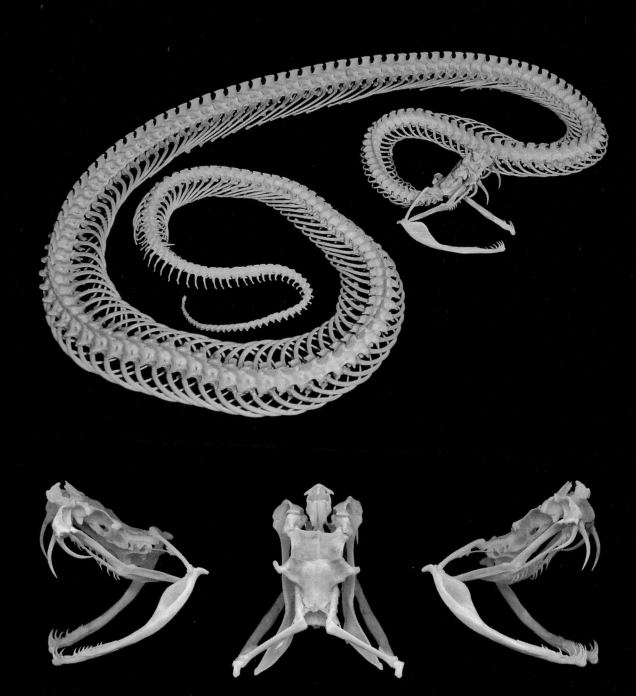

The yellowtail coris wrasse, *Coris gaimard*, is a diurnal (active during the light) predator that hunts by day and hides in a reef or the sand at night. This fish inhabits coastal lagoons, coral reefs, and rubble piles throughout the Indo-Pacific in water up to 150 feet deep, where it hunts clams, mussels, snails, crabs, urchins, worms, and brittle stars. It plucks these morsels from the bottom with long canines and crushes them with pharyngeal jaws in its throat.

Yellowtail coris wrasses, like other wrasse species, begin life as larval fish that metamorphose into juveniles with no assigned sex. As adolescents they become female, developing a new color pattern and ovaries. Numerous females are guarded by an older, larger male that protects his harem for breeding privileges. If the male gets eaten or dies due to old age, the alpha female in the harem will quickly convert to a male by developing testes and an entirely new color pattern. *She* becomes a *he* and takes over the responsibilities of harem protection and breeding.

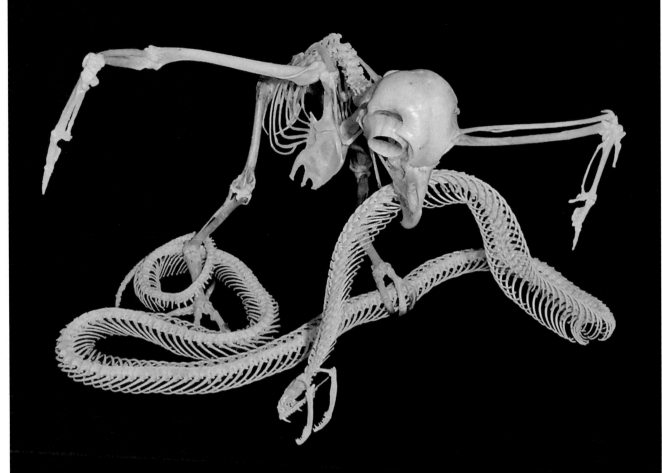

The Philippine eagle-owl, *Bubo philippensis*, is found in the lowland forests of a small number of Philippine islands. It favors habitats near water, which it uses to hunt for small mammals, amphibians, lizards, and snakes. With a wingspan up to 4 feet, it is the largest owl species in the Philippines and one of the largest owls in the world.

Philippine eagle-owls have large, powerful talons that they use to dispatch prey captured during one of their silent, stealthy attacks. They are mainly crepuscular (active as dusk and dawn) hunters that swoop down off a branch to land on their prey with a forceful barrage of claws and beak. The one pictured here has captured, killed, and begun eating a Philippine cobra, *Naja philippinensis*.

The Philippine cobra also inhabits forests found near water, as well as open fields, grasslands, jungles, farms, and some urban areas. It grows up to 6 feet long and is considered one of the top three deadliest snakes in the world.

Philippines cobras are spitters capable of ejecting venom up to 10 feet away. Their fixed fangs have small holes in the front of the teeth, rather than long grooves like non-spitters. This tiny opening allows the venom to be directed forcefully at a target, usually the eyes. The venom is highly neurotoxic and is often fatal.

■ Volitan lionfish, *Pterois volitans*, are an absolute menace to the western Atlantic, where they were introduced by people not willing to properly dispose of their pet lionfish from aquaria. Breeding populations of this species, and other aquarium fish, now exist from North Carolina to Bermuda.

Endemic to the Indo-Pacific, where they have natural predators to keep their populations in check, the lionfish along the U.S. coast are wreaking havoc on the natural order of things. They compete directly with desirable food fish such as groupers and snappers, causing their populations to decline.

Lionfish swim around our reefs with impunity because they are venomous. Their flamboyant colors and long, filamentous pectoral fins serve as a warning to potential predators that they carry a toxin. Their long, sharp, dorsal spines have venom glands within grooves along the spines that deliver a painful toxin capable of causing extreme pain, swelling, and potentially respiratory distress.

While they pose a potential health risk to anyone handling them, their meat is light and flaky and greatly resembles that of grouper. A market for their tasty fillets is growing quickly in places such as the Florida Keys.

■

138

■ The second-largest crocodiles alive today, Nile crocodiles, *Crocodylus niloticus,* are probably responsible for more human deaths than all other crocodilian species combined. This is undoubtedly due to their large home range through dozens of African nations where cities, and thus humans, are centered around water sources. They are found in lakes, rivers, streams, swamps, and even brackish water near the coasts.

Nile crocs are known to have one of the most powerful bites on the planet, and potentially of all time. While their bite *force* is high, their bite *pressure* centered on those 64 to 68 peg-like teeth is exceeded by numerous smaller animals alive today. Crocodile bite forces are used to subdue large, struggling mammals and to tear big, fleshy pieces from a carcass. Mostly, however, they feed on fish and other small vertebrates, such as turtles. When stalking prey along a river edge, their eyes are perfectly positioned to barely break the surface of the water while the rest of the body is hidden below. After creeping close to their unwitting prey, Nile crocs launch an explosive attack and snap their powerful jaws closed on their meal, dragging it back to the water to be drowned and consumed.

■

■ The cownose ray, *Rhinoptera bonasus*, has a name derived from the Greek for "nose-winged bison," quite fitting because of the blunt facial profile, large pectoral fins, and brown color. Also much like bison, this species usually occurs in large groups and is known to make long migrations, only they do so in the western and eastern Atlantic Ocean.

Cownose rays are quite majestic as they swim, with their pectoral fins appearing as wings designed for underwater flight. They are unlike other rays in that cownose rays seldom rest on the bottom. As such, they rarely pose a risk to waders or swimmers. Most rays only inject their stinger when stepped upon; since cownose rays are infrequently on the bottom, they never get stepped on.

Like all sharks, skates, and rays, their skeleton is composed entirely of cartilage. Cartilage is not as strong as bone, but it is much lighter. Without a swim bladder, these kinds of fishes need the lightest load possible to stay up in the water. Cownose rays eat very hard prey, such as clams, oysters, crabs, lobsters, snails, shrimp, and some fishes. They have a suite of flat, plate-like teeth in their mouth that they use to pulverize shells. With the mouth positioned under the head, they can swim over their food, suck it up into their mouth, and crush it to bits.

■

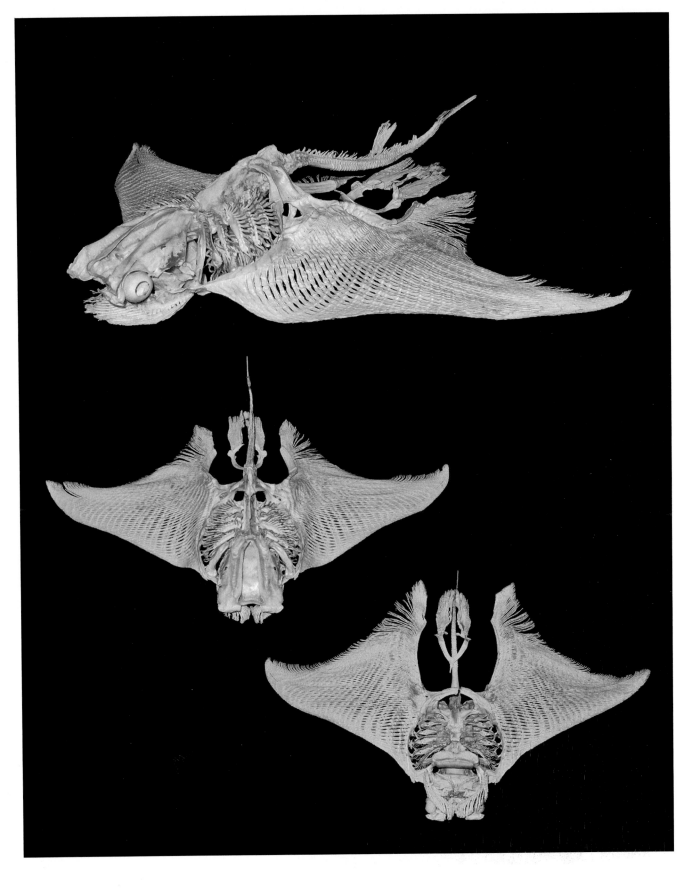

■ Also known as the lipstick tang, the orangespine tang or *Naso lituratus* is a tropical, marine fish common throughout the western and central Pacific Ocean. It inhabits coral reefs and rubble in water up to 300 feet deep. This species commonly occurs in small, tightly grouped schools where they move about the reef in search of their food, which consists almost entirely of algae. Their scissor-like jaws and rake-like teeth are perfect for mowing through vegetation.

The orangespine tang is a member of the surgeonfish family, Acanthuridae. "Acantho" is Greek for "thorn-like," and they were given this name because of the scalpel-like spines on the trunk of their tail. The spines are used in battles with rivals where the fish line up in opposite directions, side by side, and slash each other with their spines to establish dominance and secure territory or mates. Battles are rarely fatal, but significant tissue damage can occur, which is why surgeonfish skin is some of the toughest around.

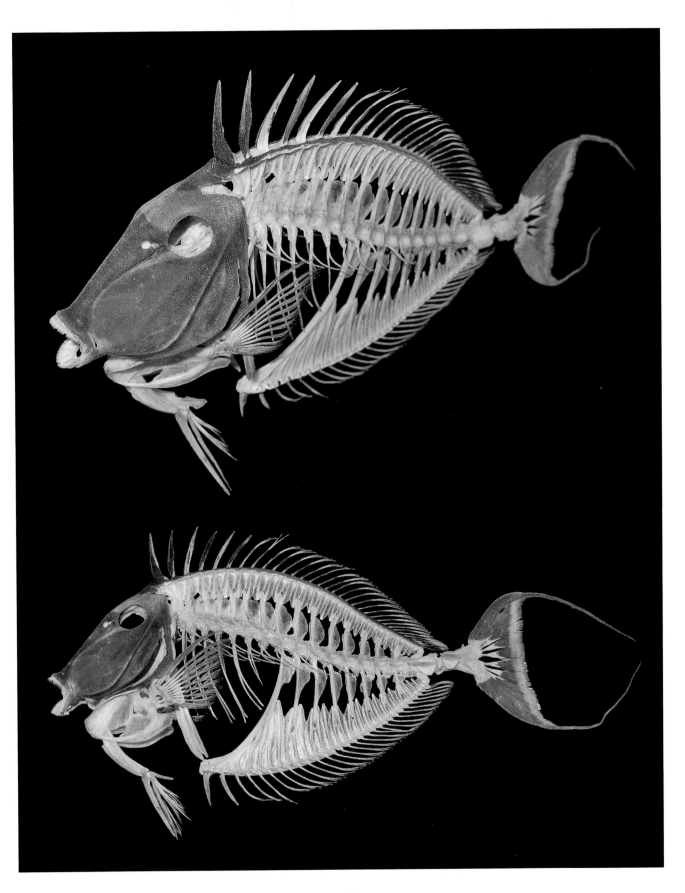

The puff adder, *Bitis arietans*, gets its common name from the loud, angry hiss it produces when threatened. Even the enormous mammals found in the puff adder's home range can hear this hiss, and they usually try to avoid it.

Puff adders are found throughout sub-Saharan Africa, as well as in Morocco and parts of Arabia. They live in every imaginable habitat, except rainforests and deserts, and are imposing predators where they're found. Wide-bodied snakes, puff adders can be up to 6 feet long and have one of the fastest strikes known. Thankfully, they are not aggressive; the large number of human bites each year (more than any other African snake) is a result of their willingness to live close to villages (where they find lots of rodents) and their excellent camouflage, leading to accidental steppings.

As with other members of the viper family, the fangs are mobile, meaning they can rotate forward to pierce their prey or any predator that threatens them. The very long fangs deliver different kinds of venom, depending on the age of the snake: cytotoxic venom when feeding on rodents as an adult and neurotoxic venom when feeding on lizards as a juvenile. The venom is also a great defense against honey badgers, warthogs, and raptors.

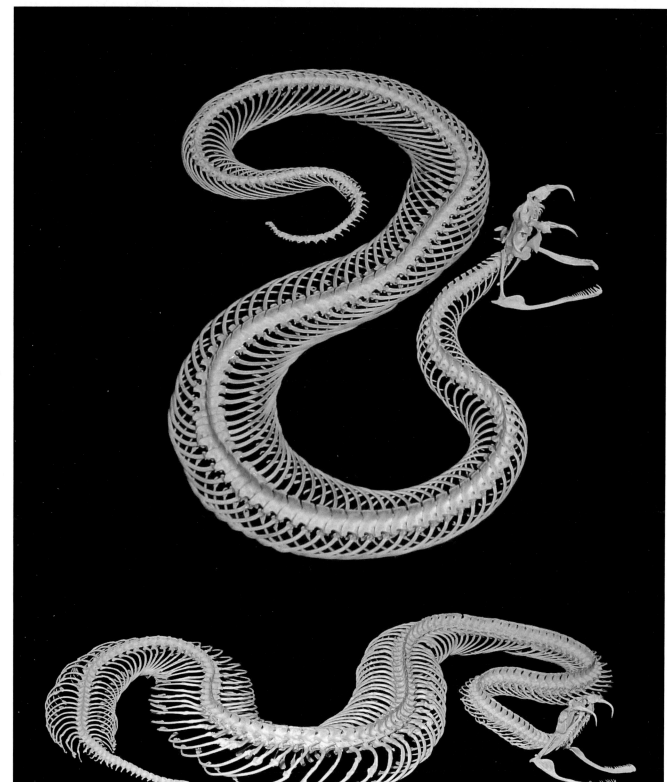

The ultimate biters, redeye piranhas, or *Serrasalmus rhombeus,* have some of the most vigorous jaws on the planet. In fact, recent research suggests that piranhas have one of the strongest bites of any animal that has ever lived (when adjusted for body size). Their stocky jaws, huge jaw muscles, biomechanical design, and extremely sharp teeth can tear through just about anything they can get their mouth over.

Also called black piranhas, redeye piranhas are more solitary than many of their marauding cousins that swim in schools, somewhat resembling packs of wild dogs. Other species, such as red-bellied piranhas, *Pygocentrus nattereri,* often take bites out of the fins of their school-mates to curb their hunger between meals.

Luckily for them, they have some of the fastest tissue regeneration of any vertebrate; bites can disappear in just 48 hours. The solitary lifestyle of redeye piranhas is altered only during the breeding season and when some unfortunate prey attracts their attention in the water.

Attacks on humans are extremely rare. In fact, most human bites are the result of human error, such as swimming downstream from a blood source or taking a piranha off a fishing hook. All piranhas do an important job of maintaining the health of other species by thinning out the weak, much like sharks do in the oceans. As such, they are a critical part of the Amazon rainforest.

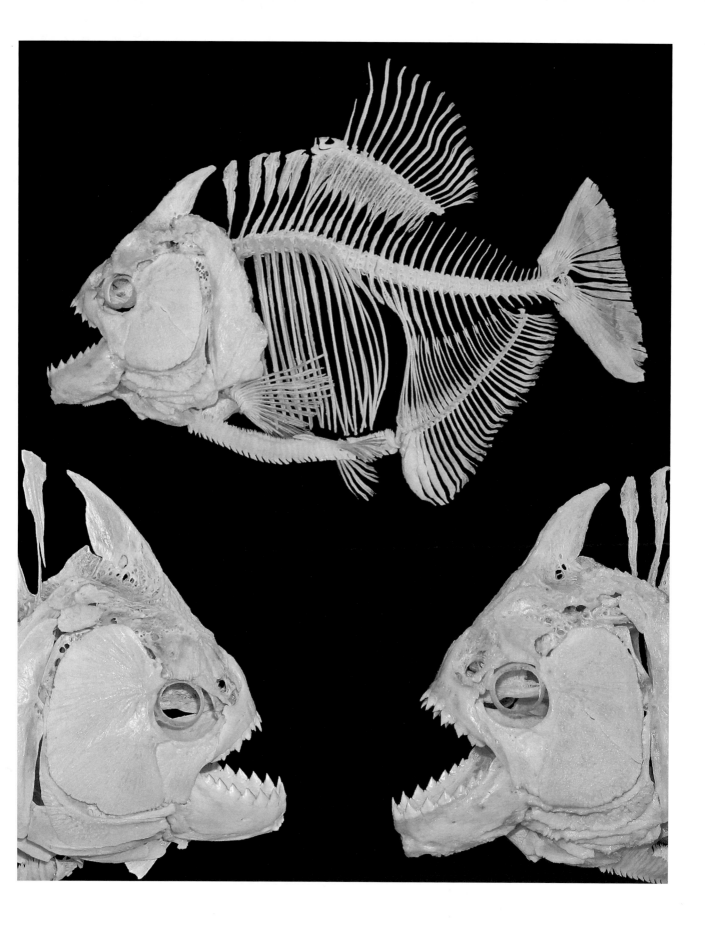

■ *Naja sumatrana*, or the Sumatran spitting cobra, is a 5-foot, venomous species of snake found in southern Thailand, peninsular Malaysia, Sumatra, and Borneo. It is native to tropical forests at elevations from sea level up to 5,000 feet. Development by humans has increased the number of interactions in places such as rice fields and oil and rubber plantations, where the snakes are often viewed as a nuisance and are killed by fearful workers. They are attracted to these locations because of the increased rodent populations that follow human development.

Sumatran spitting cobras have a highly potent neurotoxin in their venom. The venom is employed to subdue mostly rodents and frogs, but they also feed on lizards and other snakes. Once the venom is incorporated into the bloodstream of their meal, it shuts down muscles such as the diaphragm and the heart, causing death.

Spitters also have the ability to use their venom to deter potential threats. Along with the hissing and flaring of the hood also used by other cobras, spitters have modified fangs with a small hole in the front of each. This hole is pointed directly forward and can be used to spray venom yards away. The snake targets the eyes because that's the best way to get the venom into the bloodstream of their assailant. Death is possible; pain and blindness are guaranteed.

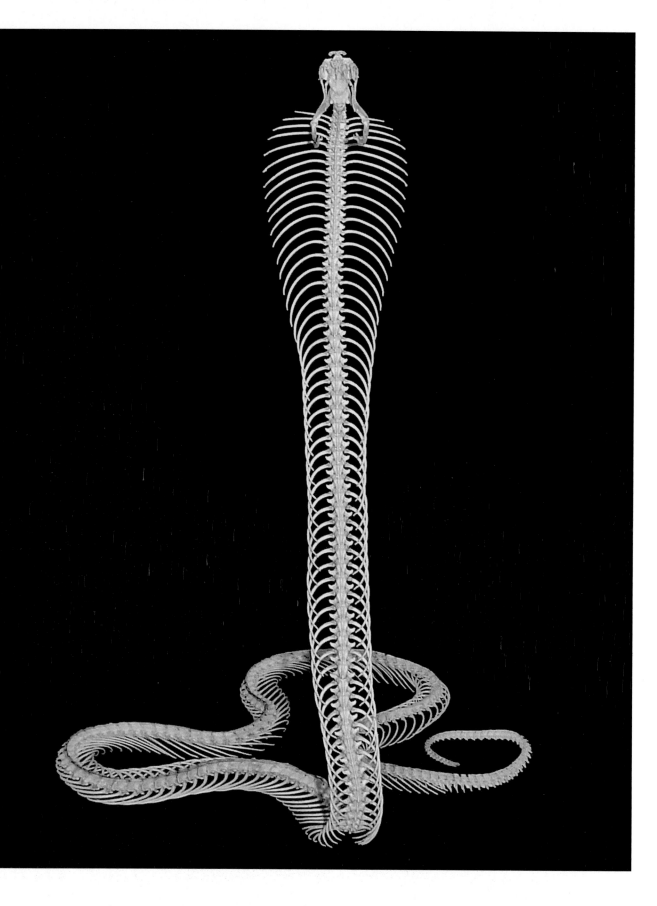

■ *Tupinambis rufescens*, or the Argentinean red tegu, is the largest of the tegu lizards. It can grow up to 4.5 feet long and weigh up to 25 pounds. Found in arid regions of Argentina, Bolivia, and Paraguay, it demonstrates little fear of humans, often living near villages. This species spends its nights in burrows and emerges during the day to hunt for food.

Red tegus are aggressive feeders and use their forked tongue to track down prey such as eggs, fruits, insects, and rodents, which they capture with powerful jaws lined with cone-shaped teeth for piercing and molars for crushing. Forceful jaws allow them to gnaw on their prey to soften it for swallowing whole.

This species has become popular in the reptile trade and can make docile pets. However, when threatened or agitated, they lift their head high, huff and puff, and flare their mouth open as a warning. Their sharp claws and whip-like tail can be used to lacerate animals they see as a threat.

■

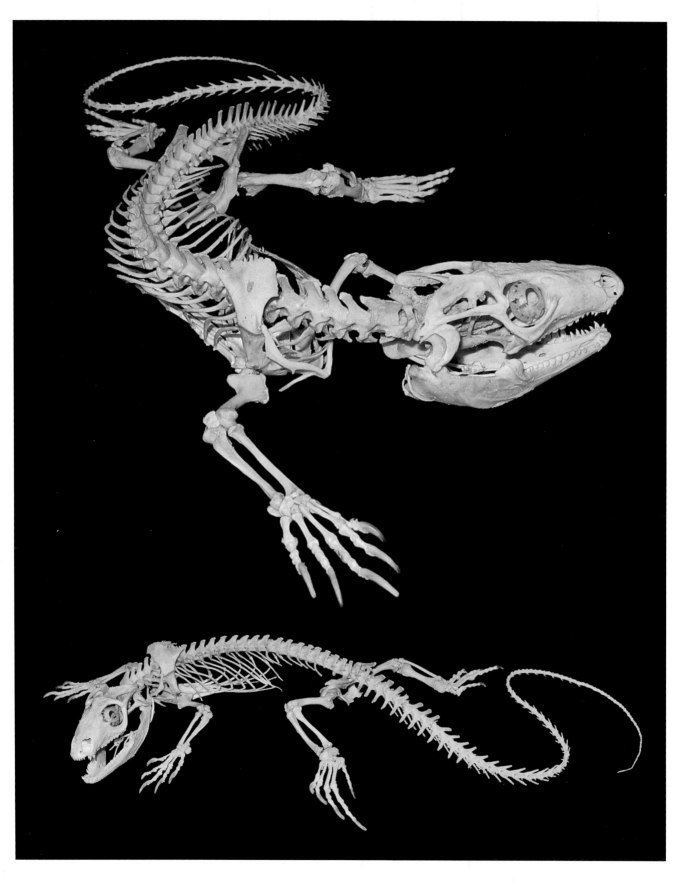

■ The king mackerel, *Scomberomorus cavalla*, is a tropical/subtropical species of marine fish typically found from North Carolina south to Brazil. It is most commonly seen along the coasts in water less than 150 feet deep. However, kings, as they're known, may also be caught near the Gulf Stream in 600 feet of water and are prized by anglers and spear fishers alike.

King mackerel are one of the fastest fishes in the sea. Their bodies are built for speed in the following ways. They are perfectly honed to a point at their snout. Their tall tail has a shallow depth, which means it has a large aspect ratio, perfect for generating powerful thrust. Their first dorsal fin lies flat in a groove on top of the fish, and the pectoral fins on the sides lie in pockets when the fish is moving at high speed; these fins are used for maneuvering and only create drag when the fish is swimming fast, so nature has selected for ways to hide them when it's time to accelerate to their maximum speed of around 50 miles per hour.

Kings feed on squid and a variety of other fishes and are voracious predators. They eat menhaden, sardines, anchovies, cutlassfish, small jacks, blue runners, and more. Their jaws rotate open when attacking prey, revealing the row of perfectly spaced, spade-shaped, razor-sharp teeth that lines the entirety of both jaws. Upon hitting another fish with these teeth at 50 miles per hour, king mackerel often cut their prey in half, which makes for easier swallowing.

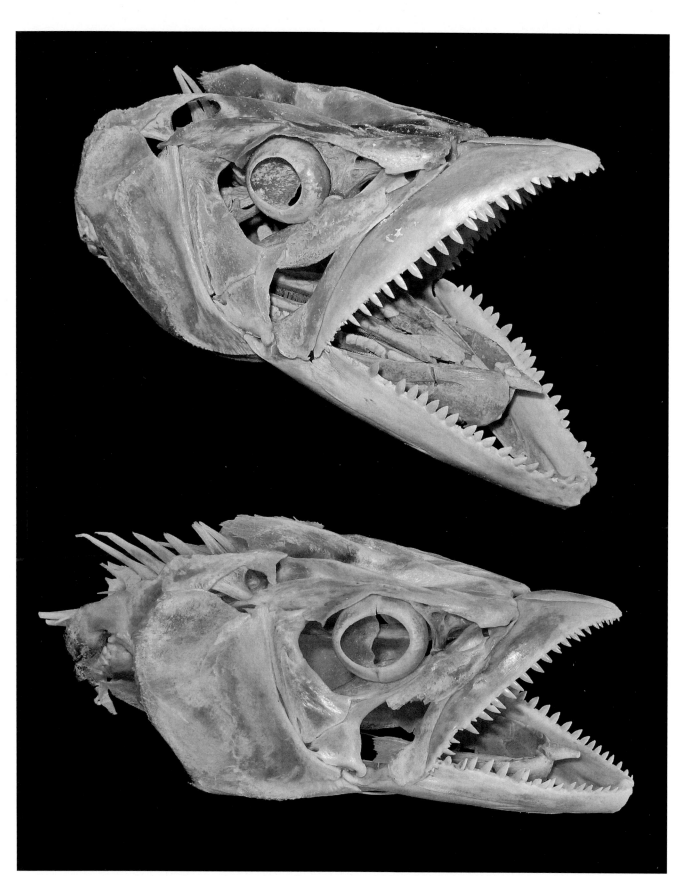

■ The eastern long-necked turtle, *Chelodina longicollis*, is found throughout southeastern Australia in freshwater rivers, lakes, and wetlands. Its other common name, "stinker," comes from the pungent odor it emits from its musk glands when it feels threatened; it can even eject this stinky fluid as far as 2 feet. Like other turtles, they hide in their shell when threatened, but their long necks, up to 60% of their shell length, do not allow them to pull their head back into the shell like most turtles. Instead, they curl their head around to one side and tuck it into the shell sideways.

Eastern long-necks are carnivorous and eat anything they can fit in their mouth, including insects, worms, crustaceans, snails, tadpoles, frogs, and fish. They tend to lie in wait or slowly pursue off-guard prey and quickly thrust their neck out when they strike. Their enlarged hyoid bones along the throat allow for rapid expansion of the neck to create significant suction during feeding.

■ Very distinctive in its appearance, the striped burrfish, *Chilomycterus schoepfii*, is a spiny fish from the western Atlantic. It inhabits seagrass beds in shallow bays and lagoons where salinities can range from 7 to 47 parts per thousand.

Striped burrfish grow up to 10 inches long and are covered with short, sharp spines that are tetrahedron shaped. The spines stand erect at all times and create an impenetrable latticework of armor for protection. Burrfish look quite cumbersome when swimming because their shield of spines keeps them from bending their bodies. Rather, they swim by coordinated action of their pectoral, dorsal, anal, and caudal fins.

Striped burrfish are members of the family Diodontidae, which translates to "two teeth." They have single upper and lower dental plates that are extremely robust. These plates are used to crush hard prey, such as crabs, mussels, clams, oysters, and urchins. Burrfish then "spit and suck" the pieces in and out of their mouth to separate the soft flesh from the hard shells.

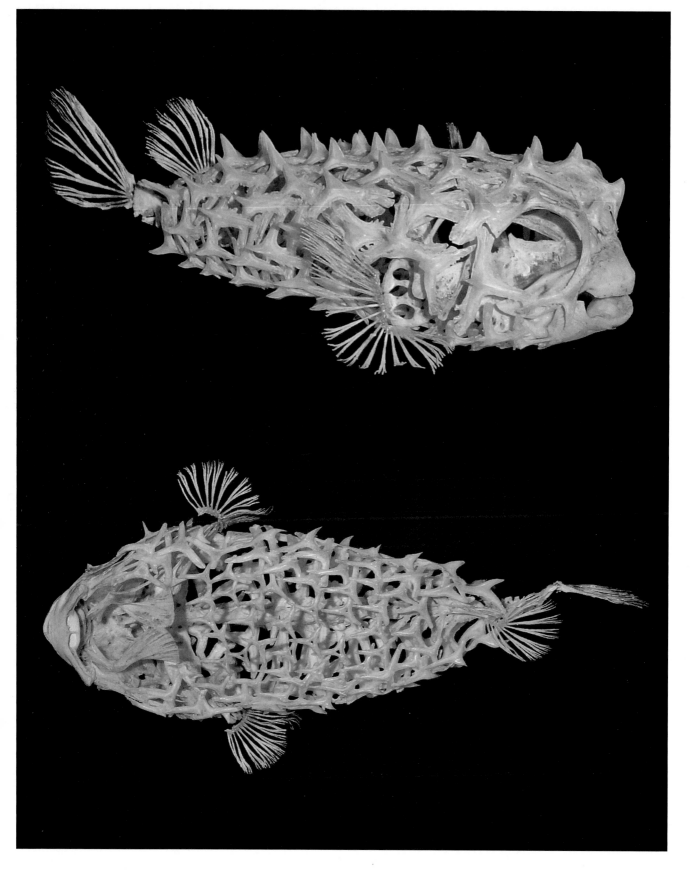

The cottonmouth, *Agkistrodon piscivorus*, is a thick-bodied, semi-aquatic, venomous snake endemic to the southeastern United States. It can be found in any type of fresh water and is even known to visit brackish (slightly salty) water on occasion. The species often basks on branches and exposed logs along waterways, contributing to their frequent encounters with humans.

Cottonmouths get their name from the stark white color inside their mouths, which they expose during hissing at potential threats. Their venom can be lethal to humans but is rarely so. It is used to subdue their prey, which includes both terrestrial animals (small mammals and birds) and aquatic creatures (snakes, frogs, and even fish). This specimen was posed with a recently captured longear sunfish, *Lepomis megalotis*, in its gut. Ill-tempered snakes, cottonmouths are known to stand their ground when threatened. I once witnessed a 4-foot cottonmouth on a levee in Florida stand its ground against my full-size truck, pulling a trailered boat.

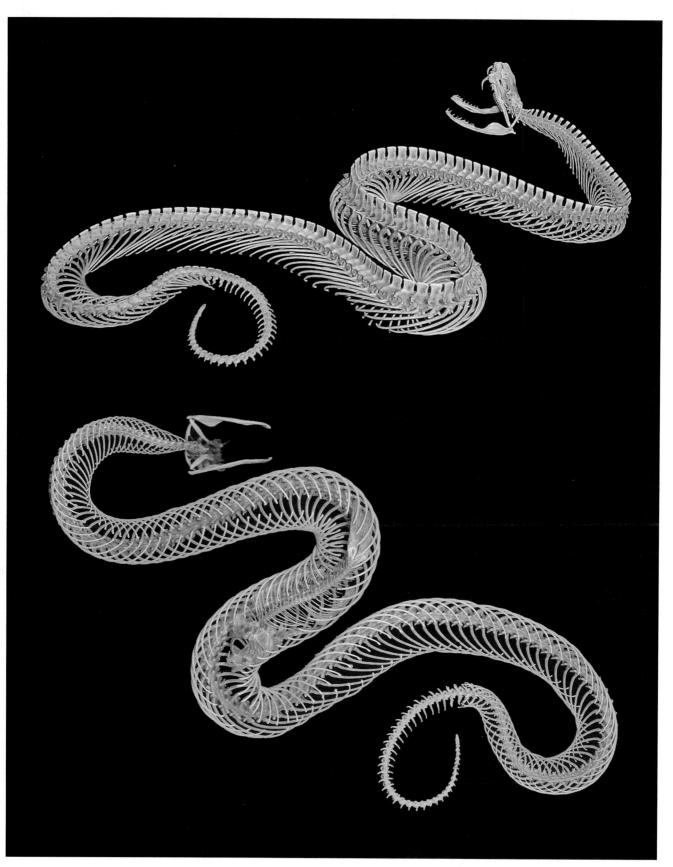

The common snapping turtle, *Chelydra serpentina*, is a large, freshwater turtle growing up to 20 inches long and in excess of 50 pounds. It occurs throughout the central and eastern United States, thrives in nearly all freshwater habitats, from streams to ponds to swamps, and is even found in brackish water.

Common snappers have a very long neck (hence their species name, "serpentina"), which allows them to move their powerful jaws and sharp beak closer to unwary prey. They're omnivores willing to eat anything they can grab, including plant material, crawdads, fish, such as this *Pterygoplichthys multiradiatus*, frogs, snakes, baby ducks, and small mammals. There are even reports of snapping turtles biting off the teats of cattle cooling off in farm ponds.

Common snappers are quick to quietly slip away to safety when encountered by humans near or in water. However, when approached on land where they are out of their element, they can be highly aggressive. They'll hiss, flare their mouth open, and snap their jaws at any threat. They also spring their neck out and push off with their hind legs, creating the appearance of "jumping" at danger. With jaws capable of biting through a broomstick, they can easily take a human finger.

With the broadest distribution of all the rattlesnakes, from Colombia to Argentina, the South American rattlesnake, *Crotalus durissus*, is arguably responsible for more envenomations than any other rattler species. They occur in a diversity of habitats, from woodlands to savannas to semi-deserts, and are frequent visitors to human habitations in search of garbage-loving rodents.

Like other mobile-fanged snakes, this species has the ability to swing their fangs from lying flat inside the mouth to protruding out for biting prey or threats. A suite of muscles pulls on certain bones, which in turn pushes the bones housing the teeth forward, exposing the fangs. Once bitten, the venom flows freely through hollow, hypodermic needle-like teeth that can be quite long.

South American rattlesnake venom varies widely across its range, from neurotoxic (toxic to nerve cells) effects in the south to blood and cell toxins in the north. Bites from this species have led to many human fatalities and can be extremely painful. As with any venomous animal, it is best to leave them undisturbed.

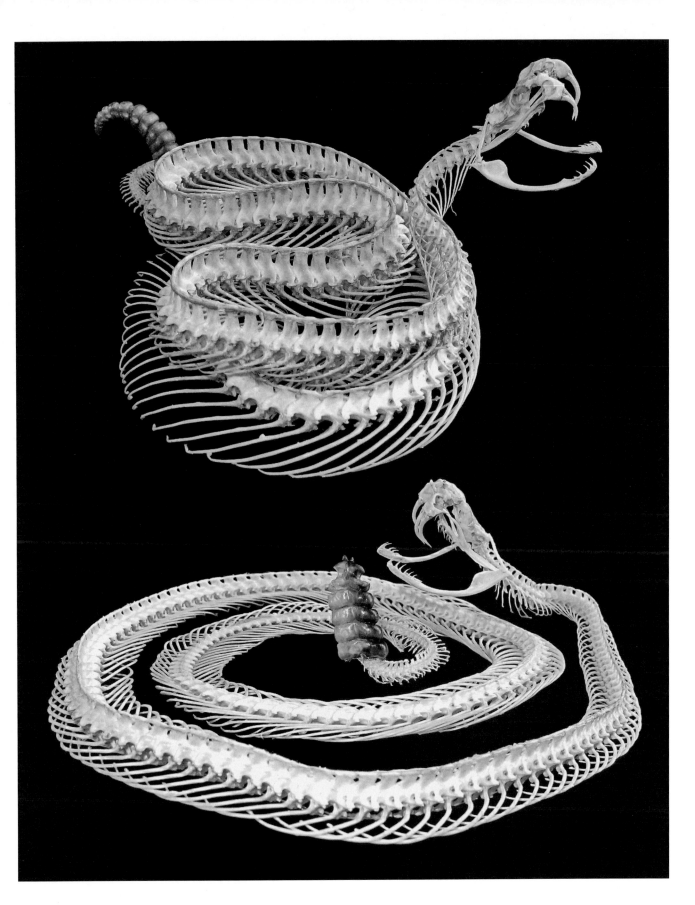

■ *Varanus exanthematicus*, or the savannah monitor, is found throughout sub-Saharan Africa. They occur in open grasslands and rocky areas where it's possible to dig burrows with their powerful legs or climb under ledges that they scale with their sharp claws. They tend to hide out during the heat of the day and are much more active during the cooler, daylight hours of dawn and dusk.

Savannah monitors eat insects, birds, eggs, rodents, amphibians, and other reptiles. Unlike many sharp-toothed monitor species, savannahs have a mix of cone-shaped and blunt, robust teeth that are useful for crushing hard prey, such as snails, dung beetles, and scorpions. Their back is lined with many bony scales called osteoderms for protection, and their tail functions as a whip. When threatened, it can easily split the skin of an assailant.

■

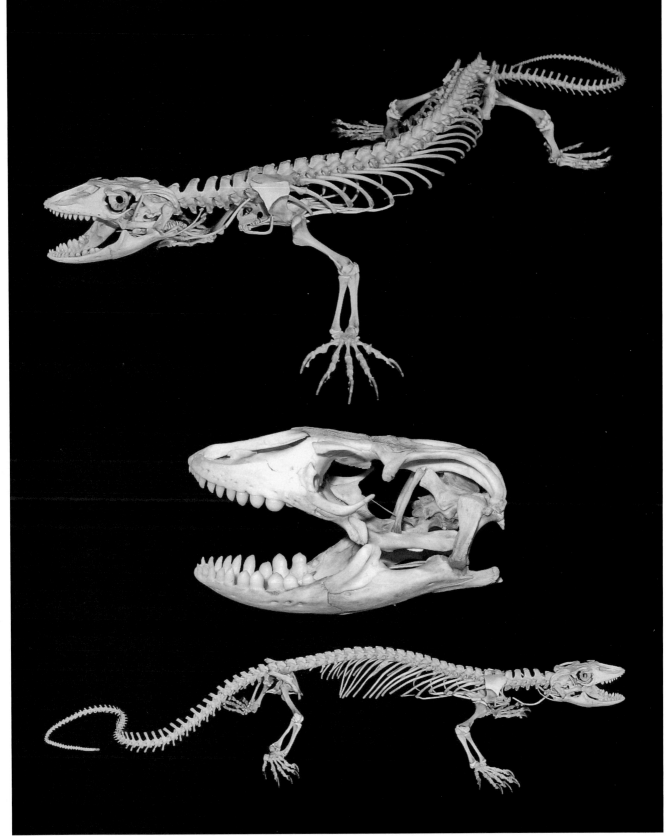

A marine species, the trumpetfish, *Aulostomus maculatus,* is found throughout the tropical waters of the western Atlantic Ocean. Its long, slender body helps it blend in with reef structures such as gorgonian or staghorn corals. Trumpetfish often sit right side up or upside down among these reef structures to hide from predators.

A long, skinny body helps them blend in well, but it makes feeding quite difficult. They have to raise their head significantly in order to blast their mouth open which, in turn, sucks in their prey. However, a long skull is tough to swing through the water quickly. So, trumpetfish have evolved an ingenious method to help swing their head—a suite of lattice-like tendons that attach their back muscles to the back of the skull. When the muscles yank, the skull pops up violently.

A trumpetfish's long mouth cavity begins as a collapsed tube that expands significantly from side to side. In the photograph opposite, you can see how it increases to four times its original size. When the skull pops up, a series of complex linkages expand the mouth cavity, which causes water, and with luck their food, to flow into their mouth.

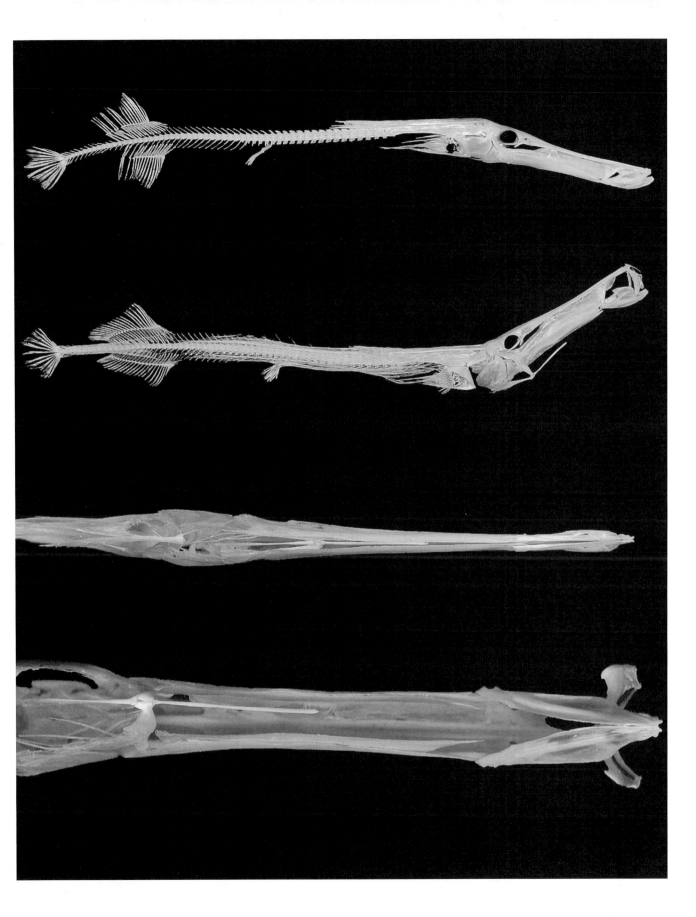

■ With its long, slender, toothy snout, the longnose gar, or *Lepisosteus osseus*, is aptly named. Its genus name means "bony scale," because its body is covered with an ancestral type of fish scale known as a ganoid scale, which feels hard, like bone. These diamond-shaped scales are loosely connected to each other, allowing for some degree of flexibility for swimming while creating an armored covering for protection from predators. A 5-foot, adult longnose gar truly only has alligators to fear.

Longnose gars live in fresh and brackish waters throughout the southeastern United States. They inhabit slow-moving streams and rivers, as well as reservoirs, lakes, sloughs, swamps, and estuaries. They can even survive in low-quality water almost devoid of oxygen because of a feature that connects their swim bladder to their pharynx. This pneumatic duct allows gars to gulp air, from which they can extract oxygen for respiration.

Gars are perfect ambush predators. They lie in wait, using their camouflage to hide among vegetation. When an unsuspecting fish swims by, their main food source is pinned between the long, toothy jaws. The gar then proceeds to make forward lunges as it loosens its grip to move the morsel closer to the mouth opening, where it swallows prey whole.

■

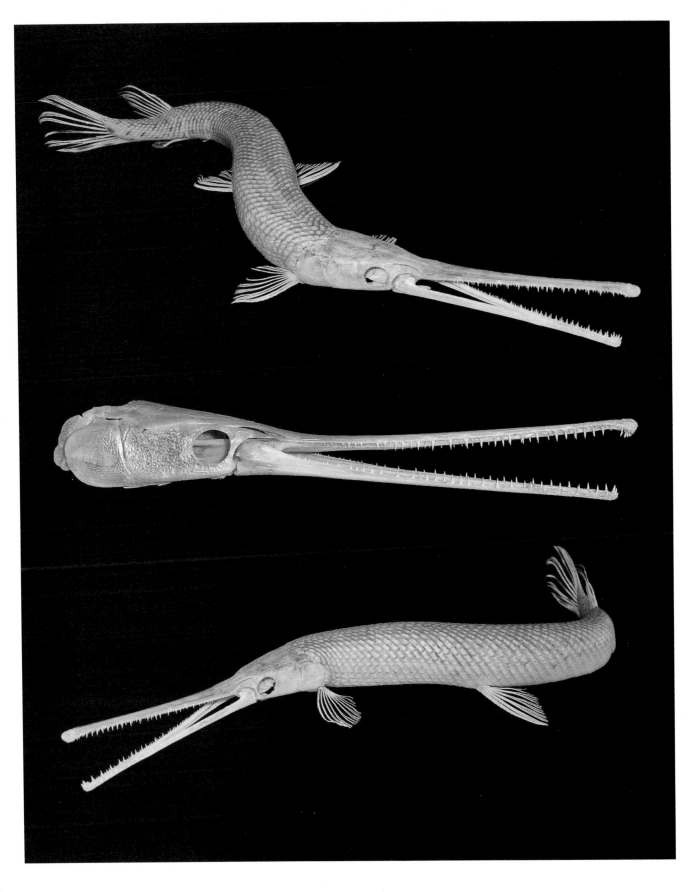

■ The ultimate diggers, eastern moles, *Scalopus aquaticus*, have huge nails at the ends of their large hands. In fact, one of the wrist bones, known as the "mole's thumb," extends out like a sixth finger to help this small mammal shovel through its subterranean home. Their chest, back, and arms are like that of an Olympic power-lifter—short, stocky, and loaded with muscle.

One way to describe the mole's method of locomotion is "swimming through solid earth." This is not far off, and the moles do this pretty much all day. To fuel this laborious lifestyle, they must satisfy their insatiable appetite. Moles have been known to eat their own body weight in worms, ants, grubs, and other underground prey each day (imagine eating approximately 150 pounds of food a day, the weight of an average person). Completely blind, moles find their food through sensing vibrations and with the aid of their senses of smell and touch. Once a prey item is located, a mole tears it into bite-sized morsels with its sturdy jaws and sharp teeth.

Eastern moles' fur is made up of extremely soft hair, so soft, in fact, that it has been used for make-up brushes. Because of its ability to repel dirt and water, their fur is also used to make fishing flies.

■

The mahi mahi, *Coryphaena hippurus*, also known as a dorado or dolphinfish, is a circumtropical, marine species known by anglers around the world for acrobatic jumps and long, line-stripping runs. The species inhabits open oceans, mainly over ledges, in water up to hundreds of feet deep. They often hang out near current edges, weed lines, and flotsam (floating wreckage of a ship or its cargo) as small as a Styrofoam cup.

Mahi get up to 7 feet long, weigh more than 80 pounds, and are voracious predators. They spend a great deal of their day darting in and out of floating *Sargassum* seaweed mats for prey such as octopi, squid, and shrimp, but they mainly eat small, pelagic (open sea) fishes, such as billfishes, jacks, tunas, and juveniles of reef species.

One feature that sets mahi apart from all other pelagic species is their pronounced skull crest, especially in males, known as bulls. The large profile of their skull creates the appearance of a square head that tapers to a slender body with a huge, powerful tail. This flat, plate-like head provides great maneuverability in the water when chasing prey or avoiding predators such as large tunas, billfishes, or sharks.

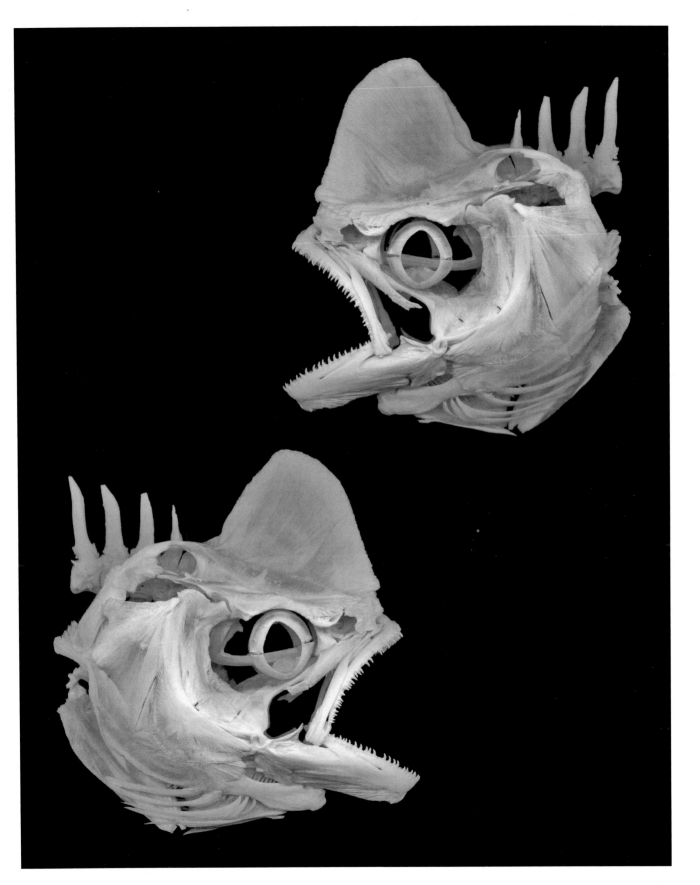

■ The devil scorpionfish, *Scorpaenopsis diabolus*, is a Hawaiian species known to natives as the nohu. Like most other members of this group, they spend their time on the bottom hiding with superb camouflage, resembling a rock or other reef structure. As such, they are often dismissed by waders, snorkelers, or divers as just another rock on the ocean floor. A poorly placed foot or hand is then pierced by the venomous dorsal spines, yielding excruciating pain.

Scorpionfish are the textbook example of ambush hunters. They lie in wait for days, sitting motionless and resembling a rock, waiting for some naive small fish or shrimp to swim too close. Because their meals are few and far between, they have one of the highest success rates of prey capture among all fishes, since they can't afford to miss an opportunity. This has led to the evolution of one of the fastest strikes known. A scorpionfish can explode its mouth open, protrude its jaws, swallow a volume of water housing prey, and close its mouth again, all in less than 20 milliseconds. The fastest human eye blink takes approximately 100 milliseconds, so these fish could theoretically feed five times in the time it takes you to blink your eye.

■ *Crotalus scutulatus*, the Mojave rattlesnake, is native to the southwestern United States and north-central Mexico. It lives in deserts, scrublands, and open woods from sea level up to 8,300 feet. The species uses cover of all types to find protection from the heat of the day, including animal burrows, stony shelves, and fallen trees, and tends to avoid densely vegetated and rocky areas, preferring open, arid habitats.

Mojaves are highly venomous pit vipers with venom containing both hemotoxic and neurotoxic components, thus affecting both blood cells and nerve cells. Their venom is considered one of the most toxic of all the snake venoms in the world and is delivered by fangs that swing forward during a strike. They employ these toxins to subdue rodents and lizards, which they ambush from hiding spots along regularly used paths. Considered aggressive, especially when cornered, Mojave rattlesnakes should be observed from a distance. The species is prone to striking, but like any animal, only does so when it feels threatened. If bitten, immediate treatment usually leads to a full recovery.

■

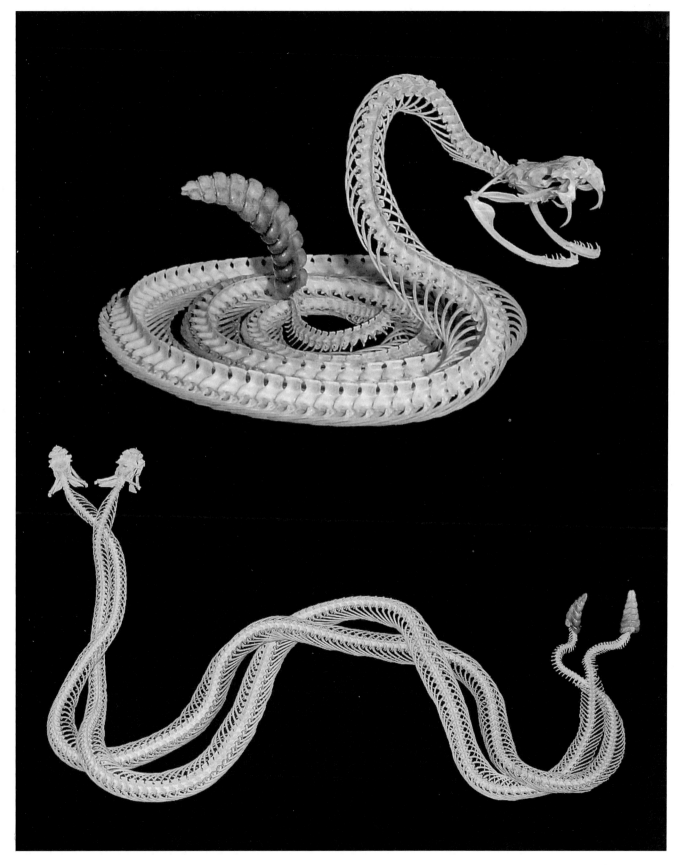

■ The wolf eel, *Anarrhichthys ocellatus*, isn't an eel at all. It gets its name from its long, eel-like appearance, which evolved so it can back into a crevice or den, where it spends most of its time, and from its toothy mouth, with teeth like a wolf.

Pacific wolf eels occur from Baja California to Kodiak Island, among other places, and live in stone piles, in reefs, and along rocky cliffs. They establish a lair and guard it against rivals, using this den as a home base for short trips out to find food.

Wolf eels have some of the most powerful jaws of any fish on the planet, and they're lined with enormous teeth. They feed on prey that most other species must ignore because of their hard shells or protective coverings, including urchins, clams, snails, mussels, abalones, crabs, and scallops. They grab food with their cone-shaped fangs, manipulate it in their mouth until they get a solid mechanical advantage on it, and utilize their huge jaw muscles to crush it with their thick, round molars.

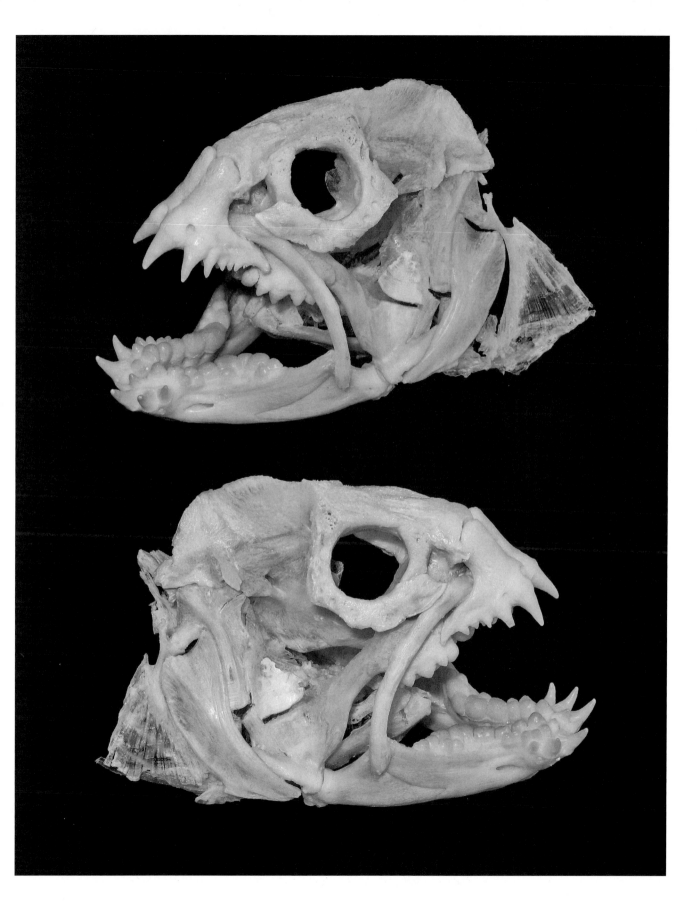

■ One of nature's most easily recognized and most feared snakes, the monocled cobra, *Naja kaouthia*, occurs in northeastern India, Bangladesh, Bhutan, southern China, Thailand, Viet Nam, and parts of Malaysia. It prefers wet habitats, such as swamps, mangroves, and paddy fields, but can also be found in grasslands, shrublands, and forests. Its pursuit of rodent prey often brings it near human habitations and agricultural lands.

Monocled cobras are identified by the striking, circular pattern on the back of their hood that resembles an eyeglass monocle. This pattern is barely visible when the snake is at rest, but when fully hooded, the skin stretches to reveal the white circle on the back, creating the impression of an eye watching behind the snake. The hood is created by the actions of specialized ribs and muscles near the head. The hood ribs are much straighter than the rest and can be spread widely to expand the neck. When coupled with loud hissing and their ability to stand one-third of their 7-foot body upright, they are quite an imposing animal.

This species is quick to respond to danger and is happy to throw rapid strikes at anything posing a threat. The venom is highly neurotoxic and can be fatal if bites are not promptly treated with antivenin medication.

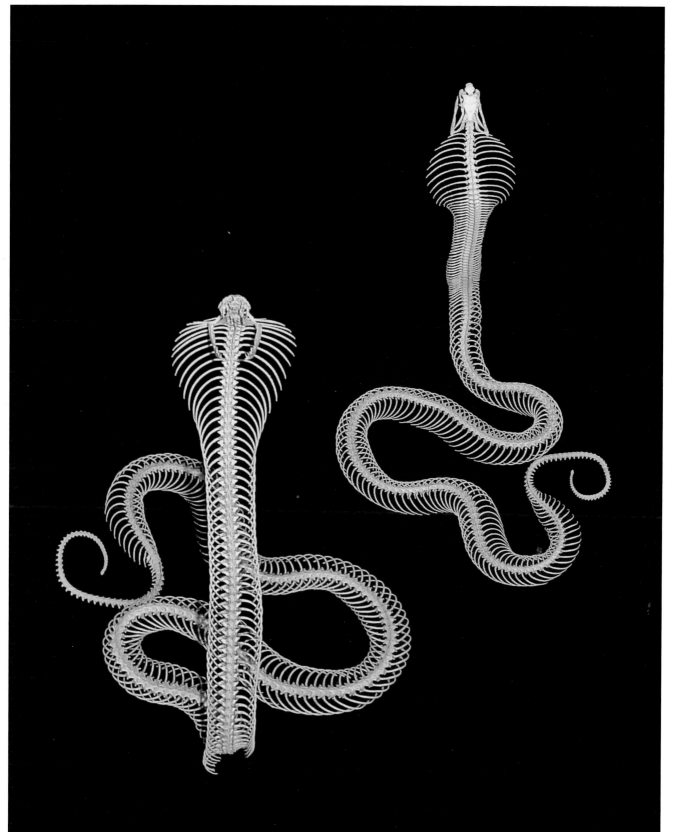

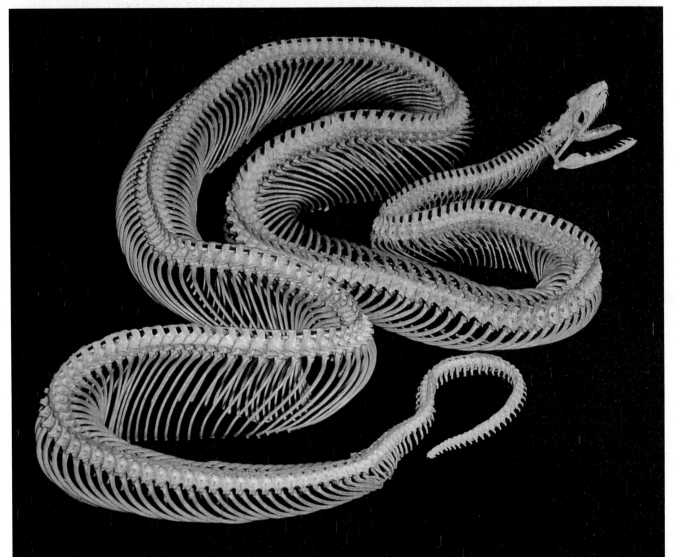

The longest species of snake alive today is the reticulated python, *Python reticulatus*. The largest ever measured was over 32 feet long and weighed 350 pounds. They're not the heaviest snakes around; green anacondas, *Eunectes murinus*, hold the title as the heaviest snake on the planet, but retics, as they're known, grow longer.

These snakes are considered extremely aggressive because of their willingness to stand their ground. When a 25-plus-foot snake decides it feels threatened, there's not much a human can do to stop it. There have been a small number of deaths attributed to this species, but these are very rare and are usually due to human error.

Reticulated pythons are found in the tropical rainforests of Southeast Asia, where they stalk along waterways for birds and mammals. Prey include rats, pangolins, porcupines, monkeys, pigs, and deer. Their meals are grabbed during an ambush attack, wrapped in ever-tightening coils, and killed. Prey is swallowed whole by flexible jaws attached to a highly kinetic skull. The top, inner rows of teeth are on moveable bones that effectively "walk" the skull over their meal.

Selene vomer, or the lookdown, is a member of the jack and pompano family, Carangidae, which is broadly distributed throughout the western Atlantic Ocean. It inhabits shallow, coastal waters and sometimes visits estuaries.

At less than 18 inches, lookdowns are one of the smaller members of their family. They tend to occur in large schools and do so because they are a favorite meal of many fish-eating predators. Their bodies are extremely flat and their skin has amazingly reflective properties, helping them use the bright sunlight in shallow water to hide from predators and prey.

Lookdowns feed on small crabs, shrimps, and fishes and have particularly protrusible jaws. The large crest on top of their head gets yanked backward by back muscles, while belly muscles pull the floor of the mouth and the lower jaw down through a series of complex linkages. This movement serves to "throw" a portion of the upper jaws outward, creating a perfectly round opening through which prey can be inhaled.

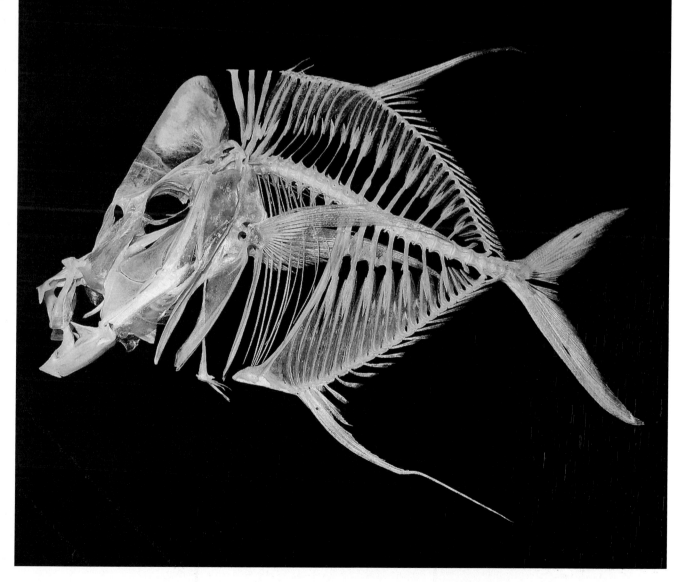

■ Alligator snapping turtles, *Macrochelys temminckii*, are the largest freshwater turtles in North America, weighing nearly 200 pounds, with shells that can be more than 2 feet across. They are found in the deeper waters of rivers, swamps, and lakes in the southeastern United States.

Alligator snappers are often confused with common snappers, *Chelydra serpentina*, yet the two differ significantly. Alligator snappers have three characteristic ridges along their carapace (shell) that resemble the ridges of the American alligator, and hence their common name. "Gator" snappers also have eyes on the sides of their heads, rather than eyes placed more toward the top of the skull like those of common snappers.

Able to hold their breath for approximately 45 minutes, alligator snapping turtles are superb lie-in-wait predators. Their shells are often overgrown with algae, helping conceal them against the bottom. They sit with their mouth wide open while their tongue wiggles like a worm, luring hungry fish in close. The jaws then snap closed with amazing speed and ferocity, focusing very powerful bite muscles along the sharp edges of their beak-like jaws.

■

186

The panther chameleon, *Furcifer pardalis*, has two bony protuberances that create a sort of elevated, horseshoe shape on its snout. Only males possess these features. Panthers are one of the largest species of chameleon, reaching over 20 inches in length, and are found on the island of Madagascar and its small, bordering islands. They live exclusively on coastal lowlands where temperatures can get up to 104°F, with humidity as high as 100%.

Panther chameleons are prized within the reptile trade because of the many different "varieties" that can be found. These varieties are a result of the broad distribution of this species within its home range, which has produced tremendous color variation, from luminous blue green with brick red spots to wine red with yellow stripes.

Like other chameleons, they have many oddball anatomical features. Their eyes can work independently, so they can look in two different directions simultaneously. Their feet have two digits on one side and three on the other, an adaptation that lets them grasp branches perfectly. Their prehensile tails give them the equivalent of a fifth set of claws. Their tongues are missiles that get shot at their food and "grab" prey with the tongue tip; the middle of the tongue holds their meal by creating suction.

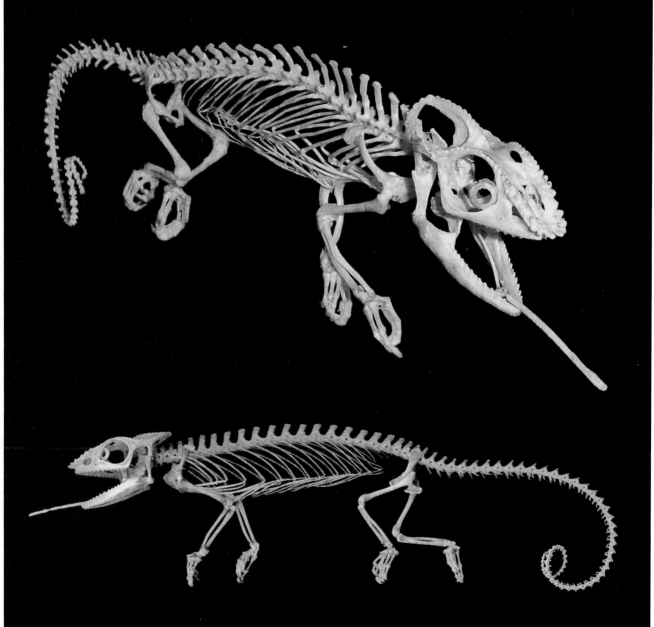

■ The four-toed hedgehog, *Atelerix albiventris*, is native to central Africa from east to west coast. It prefers grasslands and open woodlands near sea level but has been found at elevations up to 6,500 feet. It is a nocturnal animal that travels significant distances in search of mates or food during a single night, sometimes up to 5 miles. That's a long way for an animal that maxes out at 10 inches long and has decidedly short legs.

Four-toed hedgehogs are voracious predators. As small, warm-blooded mammals, they have very high metabolic rates and must consume many calories during the night to stay active. As such, not much gets ignored by hedgehogs; they eat insects, worms, frogs, toads, grubs, snails, spiders, scorpions, and some carrion.

A hedgehog's spiny fur is a protective mechanism against predators such as owls, hyenas, jackals, and honey badgers. The hedgehog will erect its spines and curl its body into a ball to protect its legs and head. It sometimes attempts to jab its spines into threatening animals, but it cannot leave spines stuck in a predator like a porcupine can.

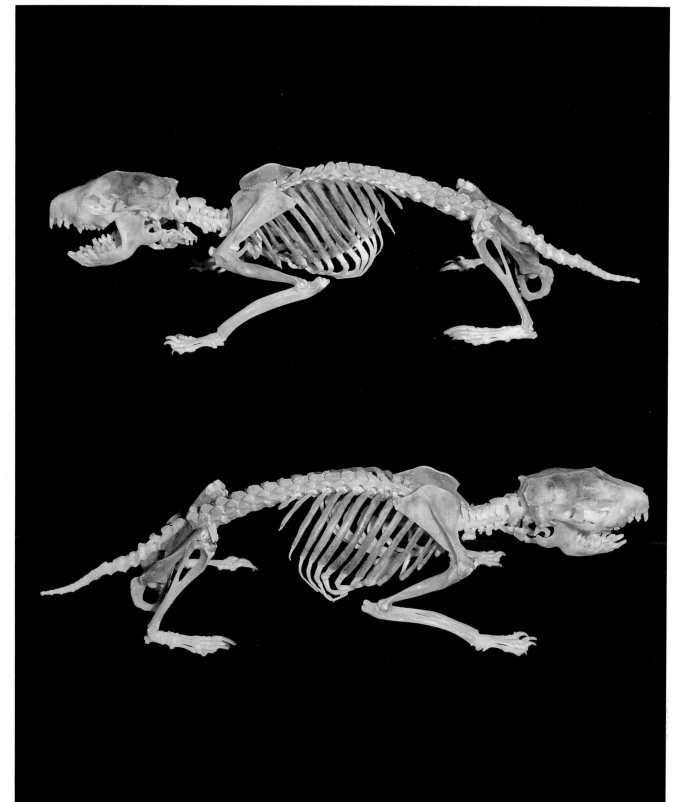

■ One of the "big five," the Burmese python, *Python bivittatus*, can reach nearly 20 feet long and weigh 200 pounds, though the average is much smaller. This species is found throughout Southeast Asia in tropical and subtropical climates. They are nocturnal and often hide in the underbrush, waiting to strike at unsuspecting prey. Burmese pythons eat small- to medium-sized mammals and birds and do so by wrapping blood flow–stopping coils around their prey, which is then consumed whole after it dies.

This species has become an exotic nuisance in South Florida, especially in the Everglades. It is suspected that captive breeding facilities damaged during Hurricane Andrew released many Burmese pythons into the wild, and they are now breeding prolifically. Their introduction is having negative effects on many native reptile, bird, and mammal species, and the state of Florida has held multiple public "round-ups" to try to curtail the population. However, with clutches of eggs numbering in the dozens, it appears that Burmese pythons are here to stay.

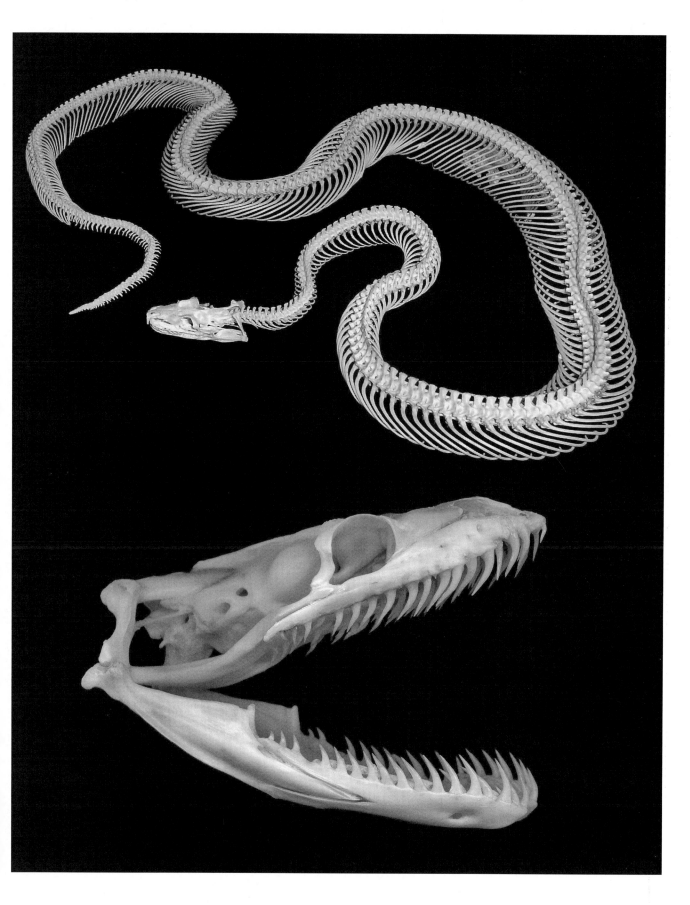

■ The round-tailed horned lizard, *Phrynosoma modestum*, is a small, 4.5-inch reptile with a very limited range within the southwestern United States and northern Mexico. It occurs in valleys, foothills, desert scrublands, and grasslands with rocky soils, at elevations up to 7,000 feet.

Horned lizards are active during mornings, evenings, and cloudy days, when they can avoid the intense sunlight. They hide under vegetation and rocks and within rodent burrows during the heat of the day and come out to hunt for ants, termites, beetles, caterpillars, and other insects when conditions are right.

When threatened, horned lizards bow their head, arch their back, and inflate their body to mimic the shape of a rock. Their coloration also helps with this impersonation. When captured, they inhale breaths of air to inflate their belly and may also employ the four spines on the back of their skull to poke their attacker.

A toothy fish, the jolthead porgy, *Calamus bajonado*, is found only in the western Atlantic, from Rhode Island to Brazil. It occurs in seagrass beds and over shallow reefs and is usually solitary.

Jolthead porgies have a diverse assortment of teeth inside their short, powerful jaws. There are cone-shaped teeth useful for capturing soft prey and molars for crushing hard prey. They feed on everything from sea urchins to clams to shrimp by pulverizing their prey into small pieces for swallowing.

Joltheads grow up to 30 inches long and weigh almost 25 pounds. They are considered excellent food fish and are highly sought after by commercial and recreational fishermen, as well as by predators such as barracuda, goliath groupers, and large snappers.

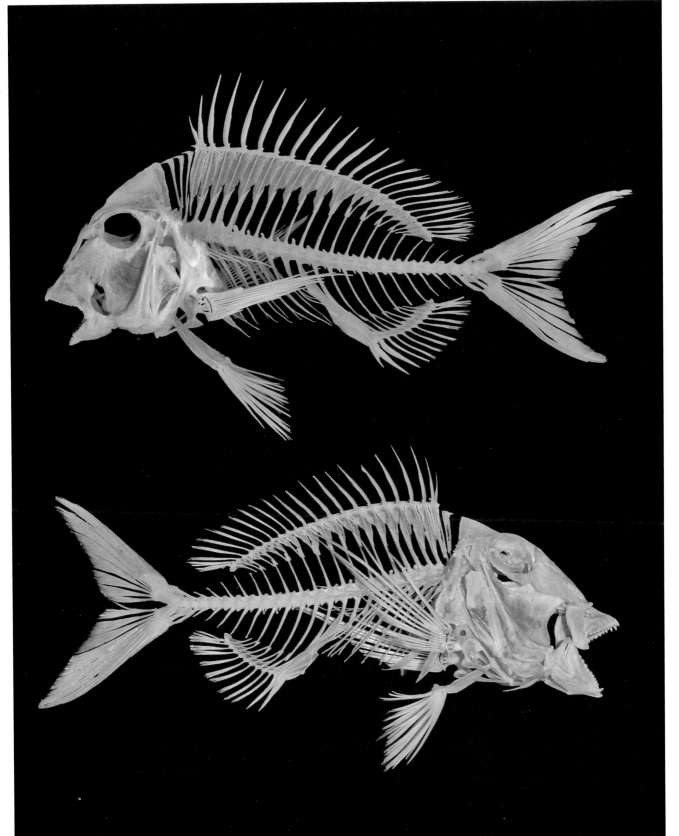

The black rat snake, *Elaphe obsoleta obsoleta*, is a very common constrictor found throughout much of the United States, from New England to Georgia and from Oklahoma to Wisconsin. It lives in a variety of habitats, from forests to prairies to swamps, and is a highly capable swimmer and climber. Black rat snakes have been observed climbing flat, brick walls and nearly smooth tree trunks. They accomplish this difficult task with significantly keeled scales along their belly that catch on any surface feature as they climb.

Black rat snakes eat mice, rats, squirrels, moles, birds, lizards, frogs, and chipmunks. They also eat many types of bird eggs, which they find after scaling tree trunks and moving through branches and trunk-holes in search of nests. They can reach 6 feet long but are decidedly slender snakes, as are most arboreal species. This helps them bridge gaps between branches without falling under their own weight.

The pair pictured opposite is locked in a mating embrace, where the male is attempting to pin the female down long enough to copulate with her. They undergo this lengthy "dance," spiraling one over the other, as the male tries to gain the advantage.

■ The woma python, *Aspidites ramsayi*, is an endangered species of snake from west-central Australia. The population has been in steep decline since the 1960s due to human development and altered land use for agriculture. They are striking to look at because of their repeated pattern of light and dark bands, extremely useful for hiding in the shadows of rocky outcroppings and ledges.

Woma pythons grow up to 7.5 feet long and are predators on small mammals, birds, and lizards. They commonly hunt their prey in burrows or underground crevices. Since they cannot constrict their prey in such tight quarters, womas are reputed to pin their prey against one of the walls of their burrow until they die.

This pair of woma pythons is in a mating embrace, where the male is trying desperately to hold the female down long enough to mate with her before another male locates them and tries to oust him. The "dance" takes on the appearance of two snakes tumbling along the sand.

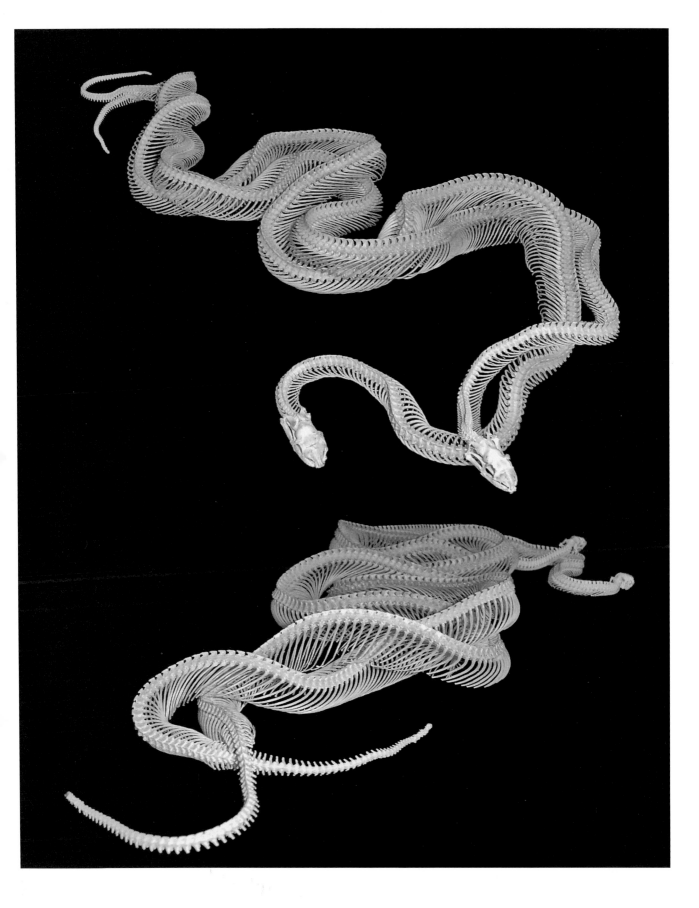

A relatively rare species of constrictor, the Boelen's python, *Morelia boeleni*, may exceed 10 feet in length. It is native only to Papua New Guinea, where it lives in cool, hilly rainforests at elevations up to 8,000 feet. Its existence is under threat from a growing human population and a desire for wild-caught Boelen's pythons in the snake trade, because successful captive breedings have been rare.

Boelen's pythons are diurnal (active during the day) and tend to bask close to their nighttime burrows in the rocks. They are ambush predators that are equally comfortable on the ground as they are in trees, waiting for unwary prey to walk, climb, or fly too close. Common prey includes frogs, lizards, birds, bats, and rodents.

Boelen's pythons are listed in the Convention on International Trade in Endangered Species appendix II, which means they are protected but not yet endangered. Their small home range, geographic overlap with humans, and desirability by snake owners do not bode well for their future.

The French angelfish, *Pomacanthus paru*, occurs along both edges of the Atlantic Ocean but is far more common in the western Atlantic from Florida to Brazil. It inhabits reefs and rocky structures in water usually less than 50 feet deep. The "acanth" portion of the genus name, which is Latin for "spine," comes from the long spine that protrudes from the preopercle (cheek bone). French angels, as they're known, may grow up to 2 feet long but are usually found in the 15- to 18-inch range.

French angelfish undergo some interesting changes throughout their lives. As juveniles, they have a diet of algae and detritus (particles of dead organisms and fecal matter) but contribute significantly to "cleaning stations" on reefs. This is where fish such as jacks, snappers, morays, and groupers stop in to be picked over for ectoparasites (organisms on the skin) that are plucked off and eaten by the cleaners. Amazingly, many of these predators would eat a juvenile French angelfish any other time, but they call a truce when visiting cleaning stations.

Adult French angels eat a diverse diet of mainly sponges but also tunicates, soft corals, and algae. They have an interesting second joint in the lower jaw. As the fish feeds, it opens its mouth and juts out its jaws toward the sponge, and the tip of the jaws are able to close like little scissors before the jaws are drawn back into the skull. The numerous, comb-like teeth are perfect for excising chunks of sponge tissue.

The barracuda of the deep, the long-snouted lancetfish, *Alepisaurus ferox*, is a circumglobal species of marine fish found in the deep waters of every ocean except the polar seas. They are often caught as bycatch by anglers and commercial boats pursuing tuna and other valuable species in water as deep as 2,500 feet.

Functionally and ecologically, they serve a role very similar to barracudas on coral reefs. Lancetfish eat other fish species, as well as squids and shrimps, and do so with gusto. They have long, tubular bodies and a powerful, forked tail, helping them move efficiently through the depths. Their musculature suggests that they cannot sustain fast speeds for long periods of time and are likely ambush predators. However, this has never been observed, since no one has ever seen them feed in the wild because of the depths at which they occur.

Long-snouted lancetfish have dentition able to slice through any type of soft prey. The tips of the jaws are each armed with four long daggers perfect for piercing prey at high speeds. The rest of the jaws are lined with dozens of extremely sharp teeth for slicing in half whatever they strike.

An iconic animal of the American southeast is the American alligator, *Alligator mississippiensis*. They are found in freshwater lakes, swamps, and rivers, with the occasional visit to local ponds and pools. The largest reptile in North America, males are rumored to reach up to 18 feet in length, though "gators" longer than 14 feet are rarely found today.

American alligators are known to have one of the strongest bites on the planet, with large males producing bite forces in excess of 2,000 pounds. That force is critical when used to capture prey such as a large deer along the shore or to crush the shell of a large, aquatic turtle, both of which are regular meals for an alligator.

While their jaws are extremely powerful for biting, their teeth lack the ability to cut through prey. Instead of chewing through their food, gators first drown their prey, then "fling" it in an attempt to snap off pieces of food. Or they will bite down and perform a "death roll" in order to twist off pieces of meat. Every bite gets swallowed whole, and every bit of the animal is consumed.

Redtail catfish, *Phractocephalus hemioliopterus*, are some of the largest fish in the Amazon. They are reputed to grow to nearly 6 feet long and 180 pounds, which makes them one of the top predators where they occur.

They inhabit large rivers, streams, lakes, and reservoirs in Colombia, Venezuela, Guyana, Ecuador, Peru, Bolivia, and Brazil. They feed on anything that fits in their mouth, including fish, crawdads, insects, worms, snakes, frogs, aquatic birds, turtles, and even small mammals such as baby capybaras.

Like other catfishes, they lack scales, and their bodies are covered with taste buds for tracking down their next meal. Redtails are members of the "long-whiskered" catfishes, so they also have six long barbels around their mouth that aid in prey detection and location. Contrary to popular belief, the whiskers of a catfish cannot sting; they are simply extensions of the skin that assist in localizing the source of what they taste in the water.

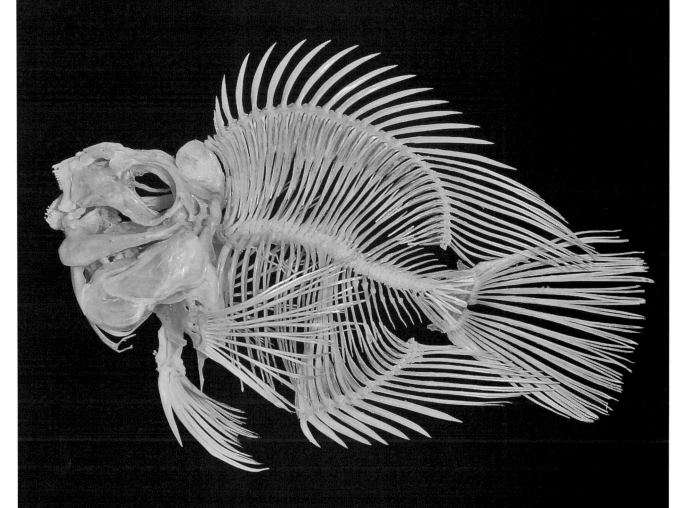

The blood parrot cichlid is a cross between a Midas cichlid, *Amphilophus citrinellus*, and a redhead cichlid, *Paraneetroplus synspilus*. Cichlids are a highly diverse family of fish found in South America and Africa and have the potential to hybridize more readily than other families. Aquarists manipulate this tendency to design new variants of cichlids for sale in the industry.

Blood parrots get their name from their vibrant red color and from their malformed, beak-like mouth. This deformity is a result of human-induced hybridization of their parent species. As hybrids, they are infertile, have misshapen mouths that do not fully close, and have numerous compressed vertebrae, among other problems. They grow to 8 inches long and are a favorite of aquarists for their color, shape, and behavior, though many collectors shun breeders that produce them for ethical reasons related to creating a fish that has difficulty feeding itself.

■ The Russian tortoise, *Agrionemys horsfieldii*, is found in Afghanistan, Pakistan, Iran, and some former Soviet territories. They inhabit dry, open landscapes of sparse grasses and bushes with sandy or clay soils. To avoid the intense, daytime sunlight, they dig burrows into the soil and are most active during crepuscular (dawn and dusk) periods. Their burrows also serve as hideouts from predators.

Digging burrows is accomplished with short, powerful legs tipped with four toes, for which they are also known as four-toed tortoises. Their toes are thicker and stronger than those of five-toed species, so they can dig much more efficiently.

Russian tortoises are relatively small and only grow up to 10 inches long. They feed on vegetation, including grasses, broad-leaf weeds, flowers, and fruits. Adapting well to captivity, they have become a favorite as household pets.

The sabre squirrelfish, *Sargocentron spiniferum*, whose name means "spiny, stinging sargo," is a marine fish found in the tropical waters of the Indo-Pacific. This is the largest species of squirrelfish and inhabits reef flats, lagoons, ledges, rubble piles, coral heads, and sandy bottoms. Here, they hide during the day and hunt at night in search of crabs, shrimps, small fishes, and cephalopods, which they locate with their enormous, light-gathering eyes.

Sabre squirrelfish are found in water up to 400 feet deep and usually move about singly or in small groups. They are common prey for many larger reef inhabitants, which drove the evolution of their namesake—the venomous, preopercular spine located on the base of the gill cover. They also possess an opercle spine, a serrated operculum bone, and numerous dorsal and anal fin spines, which pierce anything trying to eat them.

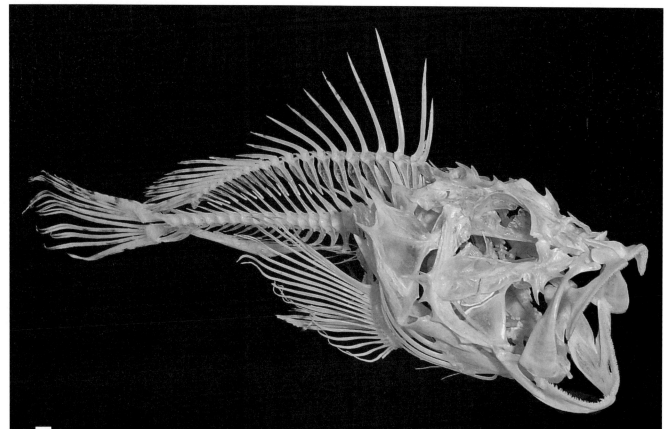

The red scorpionfish, *Scorpaena scrofa*, occurs in the temperate waters of the eastern Atlantic, including the Mediterranean and Adriatic Seas. It spends its entire life on the seabed and lives in muddy, sandy, and rocky habitats. Red scorpionfish can occur as deep as 1,600 feet but are known to visit shallow, brackish (slightly salty) waters as well. They can grow to 20 inches long.

Like other scorpionfishes, reds spend their time hiding on the bottom, with terrific camouflage resembling a rock or other reef structure. They often get missed by waders, snorkelers, or divers descending to the ocean floor, where a poorly placed foot or hand is then pierced by the venomous dorsal spines.

Red scorpionfish are ambush hunters. They hide in burrows and caves during the day and come out at night to hunt. Creeping close to prey, they slowly maneuver their way along the bottom, resembling something other than a fish with their clever camouflage. They feed on crustaceans, fishes, and mollusks with one of the fastest feeding strikes of any species.

A highly specialized fish, the sheepshead, or *Archosargus probatocephalus*, lives in the western Atlantic. It inhabits rock pilings, mangroves, worm-rock reefs, jetties, and piers from Nova Scotia to Brazil.

Sheepsheads occur wherever underwater structures become encased in barnacles, clams, oysters, crabs, or other very hard prey. They bring an array of specialized feeding adaptations to the table when it's time to eat. Sheepsheads have short, heavy jaws that are driven by large cheek muscles for powerful biting. They also have two types of teeth on their jaws—incisors up front to cut prey loose and molars in the back to pulverize them into smaller pieces.

Growing up to 30 inches long, sheepsheads can weigh as much as 22 pounds. They are prized game fish for their mild-tasting, white fillets, which also makes them a favorite of ocean-going predators, such as barracudas, goliath groupers, bull sharks, and tarpon. To deter threats, they have a suite of 17 large, sharp spines in their dorsal, anal, and pelvic fins that can make swallowing them a painful endeavor for any predator.

■ The most feared freshwater fish on the planet, red-bellied piranhas, *Pygocentrus nattereri*, have not rightfully earned that designation. Often portrayed as blood-thirsty swarms that hunt and kill anything that swims, red-bellied piranhas are actually omnivores and rarely attack people. In fact, their swarms result from the same schooling behavior used by any fish considered prey by giant otters, catfish, arapaima, and freshwater dolphins.

Red-bellied piranhas are, however, opportunistic feeders. They do a tremendous service to the local ecology by removing any weak, sick, or dying individuals. Once a target is identified, school-mates tend to aggressively "get their fair share" as quickly and ferociously as possible. This is accomplished with some of the most forceful jaws on the planet, for their body size. Once that force is focused on the tips and edges of their cleaver-sharp teeth, bite pressures can exceed thousands of pounds per square inch. Culminating with ferocious head-shakes, the teeth slice through nearly any tissue, including fruit skin, nut shells, insect carapaces, fish scales, turtle shells, caiman skin, and mammal bones.

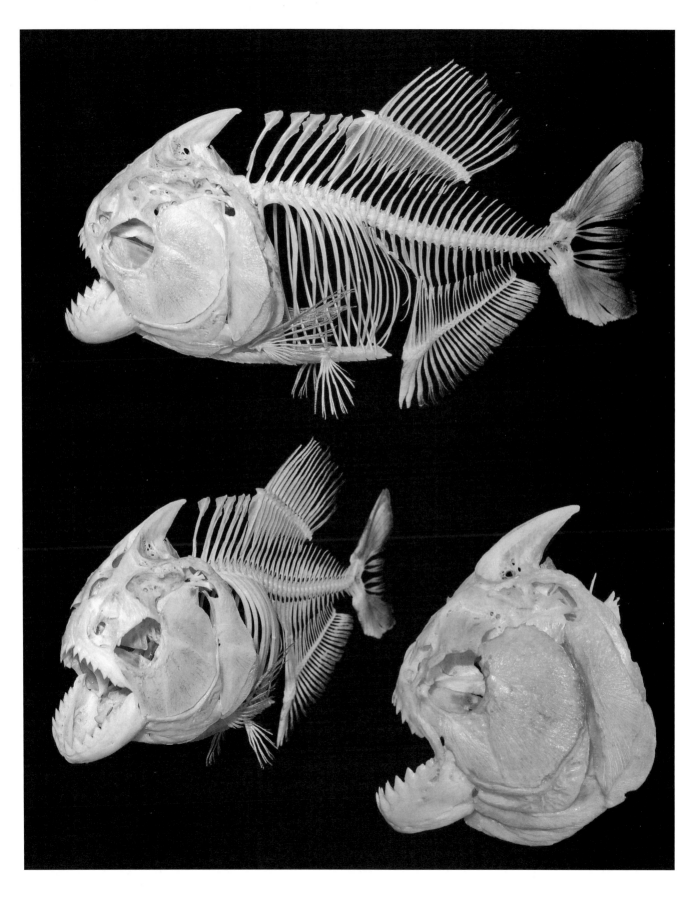

The gaboon viper, *Bitis gabonica*, has the longest fangs of any venomous species and uses them to deliver both neurotoxic (harmful to nerve cells) and hemotoxic venom. The neurotoxin shuts down its victim's ability to breathe, while the hemotoxin destroys blood cells and vessels. Gaboons use these toxins to immobilize their prey, which includes small mammals, ground-dwelling birds, lizards, and amphibians. Humans are rarely bitten, but when it happens the result can be fatal. Left untreated, a person will bleed out internally while paralysis of the diaphragm shuts down their breathing.

As with other members of the viper family, their fangs are mobile, meaning they can swing out on hinges to pierce their prey or any predator that threatens them. They do not need 2-inch fangs to feed on small prey, but in their native range of east-central Africa they have to be able to defend themselves against some large, thick-skinned threats, such as lions, elephants, rhinos, and ostriches.

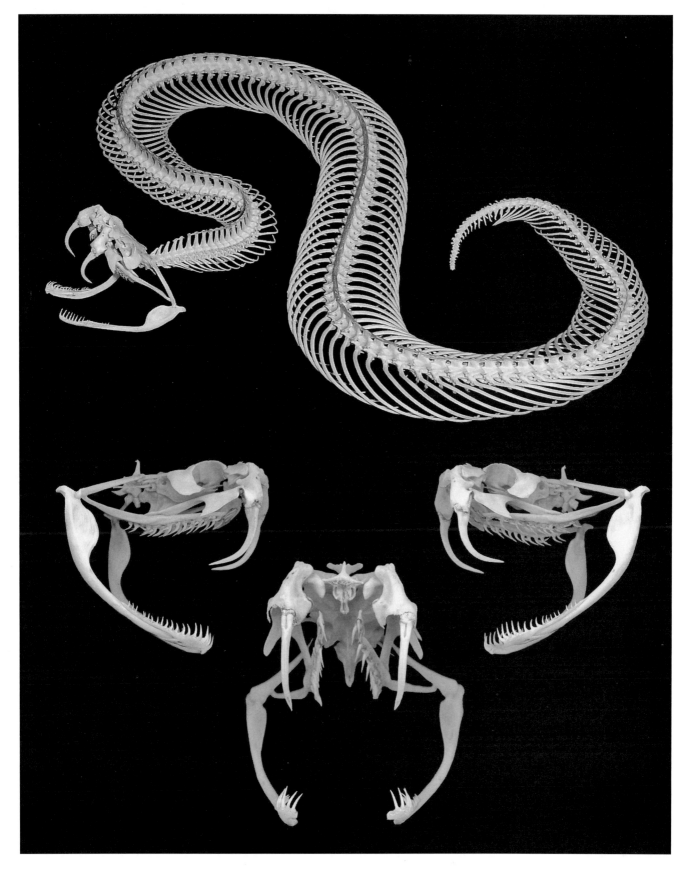

■ The houndfish, *Tylosurus crocodilus*, or crocodile needlefish, is the largest member of its family, reaching 5 feet in length and weighing more than 10 pounds. It inhabits pelagic (open ocean) realms of lagoons, reef edges, and breaks, where it stalks other fishes for food.

Houndfish have the textbook shape for high-speed swimming. Their face is tapered to a point, and their body is long and cylindrical. Also, the posterior position of the dorsal and anal fins couples them with the tail, which has a high aspect ratio, meaning that it is very tall but not very deep. This causes all three fins to provide thrust during swimming.

Hundreds of needle-like teeth in their mouths make them quite adept at holding onto struggling fishes, which they capture by high-speed pursuit and surprise dart-like attacks. In fact, this species is so dart-like that it occasionally accounts for injuries to humans, who are stabbed by houndfish beaks during their acrobatic leaps when hooked or startled.

The tripletail wrasse, *Cheilinus trilobatus*, is named for its big lips (the Greek *cheilos* means "lips") and for the three-lobed (*trilobatus*) tail observed in large adults. It inhabits coral lagoons, coastal reefs, rubble piles, and grass flats throughout the Indo-Pacific in water up to 100 feet deep. Also known as the Maori wrasse, this species is very common throughout its distribution.

They grow up to 18 inches long and have specialized teeth to capture and process a diverse assortment of hard prey, including clams, mussels, snails, crabs, and shrimps. Occasionally, they feed on smaller fishes by darting out from cover within the reef for a surprise snatch and grab. They, like all wrasses, have sophisticated pharyngeal jaws within their throat that allow them to process prey quickly and efficiently. This

second set of jaws has greater mobility and functionality than those of other, non-wrasse species of fish.

Amphilophus citrinellus, or the Midas cichlid, is a large member of the very diverse fish family Cichlidae. It is endemic to the San Juan River of Costa Rica and Nicaragua, where it is known to be extremely territorial and aggressive. This cichlid has been transplanted elsewhere and is an adept survivalist when introduced to new waters, such as the canals around Miami, Florida.

At greater than a foot in length and weighing over 2.5 pounds, Midas cichlids are legitimate threats to other species. Voracious predators of other fishes, crustaceans, insects, and mollusks, they are known to consume plant matter as well—true omnivores. They establish territories and defend them against others, even larger species of fishes. The large, bulbous region atop the skull is useful for ramming others in shows of dominance and territoriality. Their strong jaws and robust bodies allow them to bite and relocate anything from logs to rocks to other fish. Keepers often describe them as "belligerent" in tanks.

■ The water monitor, *Varanus salvator*, is a water-dependent lizard found in tropical forests, swamps, river deltas, shorelines, and even saltwater bays from India to Indonesia. It is the second-heaviest and third-longest lizard alive today. Komodo dragons, *Varanus komodoensis*, get heavier, and crocodile monitors, *Varanus salvadorii*, get longer, but water monitors can grow to an impressive 9 feet in length and reach 100 pounds. Their powerful tail functions to produce thrust in the water, and this species swims and dives with tremendous grace and speed.

Water monitors are highly aggressive toward others and spend their days alone, hunting for anything that moves. They eat birds, bird eggs, rats, monkeys, squirrels, fish, frogs, snakes, baby crocodiles, croc eggs, tortoises, sea turtle eggs, crabs, and insects; they have even been known to exhume and consume deceased humans! Their vigorous jaws and sharp teeth allow them to tear pieces off carcasses and carrion rather than swallowing prey whole, like smaller monitor species do.

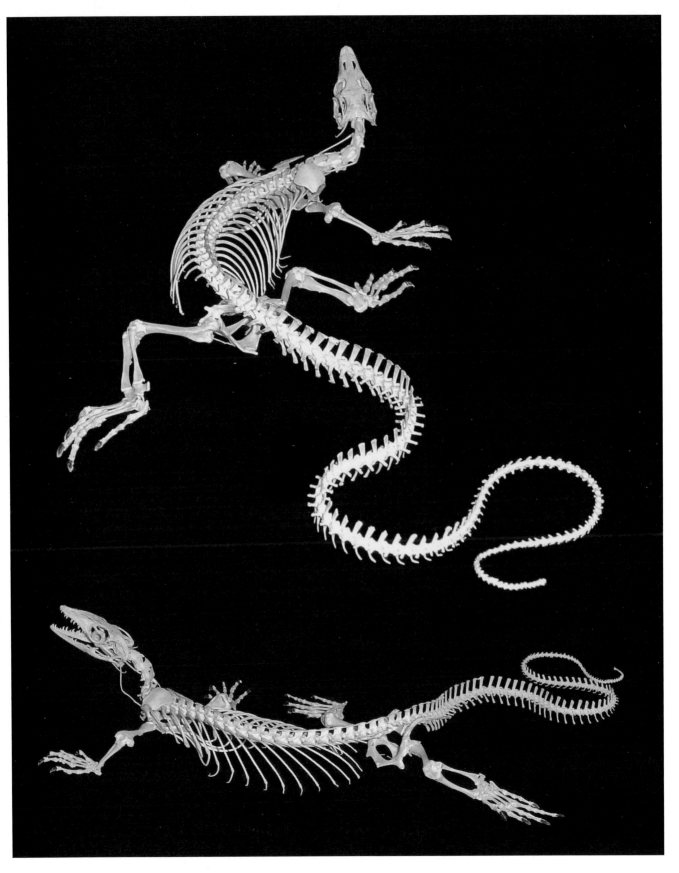

The blood python, *Python brongersmai*, is a short, fat-bodied species of constrictor that rarely grows to more than 6 feet long. The name comes from the vibrant red color pattern on the skin. Native to Thailand, Sumatra, and Malaysia, they live in forests along lowland, swampy habitats. They are prolific predators of small mammals and birds, especially in plantations where mouse and rat numbers are quite high.

Nocturnal (active during the night), blood pythons spend a great deal of time along the edges of streams or swamps, barely submerged under water. They are lie-in-wait predators that sit perfectly still for up to a week, waiting for some unsuspecting prey to venture too close. Like other pythons, they have heat-sensing pits along their snout that help them detect warm-blooded prey in the dark.

Blood pythons are one of very few reptile species known to incubate their eggs. A female will curl up around her eggs and shiver when temperatures drop, producing heat for the developing embryos.

■ From the Mediterranean Sea to the Gulf of Mexico, *Dasyatis centroura*, the roughtail stingray, occurs in tropical and warm temperate waters of the Atlantic Ocean. It inhabits muddy and sandy bottoms, where it can lie flat on the bottom and hide its edges with silt or sand.

Roughtail stingrays of all sizes, the juvenile pictured opposite included, are members of the class Chondrichthyes, which includes sharks, skates, rays, and chimaeras. This group is composed of animals that completely lack bones; their skeletons are made entirely of cartilage. Roughtail stingrays get their common name from the rows of jagged, thornlike spines that line the tail.

As a member of the stingray family, Dasyatidae, they have a highly specialized, modified scale(s) that lies on the surface of the tail. This saber has rear-facing barbs that cause it to stick and stay once plunged into a threat, such as an unwary wader that accidentally steps on a ray. The spine delivers venom that causes severe pain, swelling, and infection, though it is rarely fatal, and often has to be surgically removed.

■

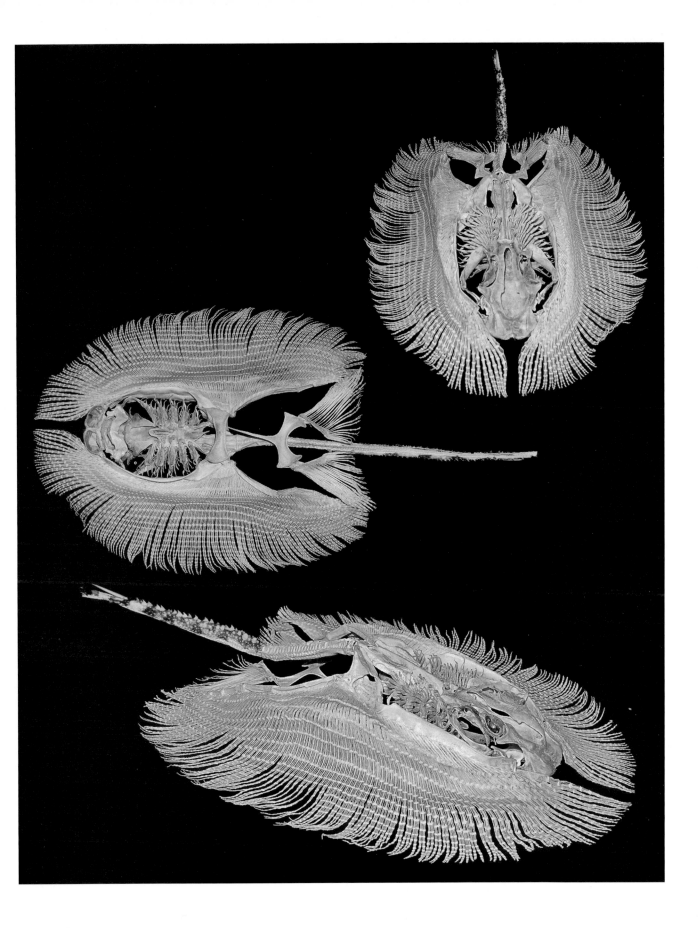

■ The black drum, *Pogonias cromis*, is the largest member of the drum and croaker family, Sciaenidae. It can grow to nearly 150 pounds and has amazing fighting strength when hooked by an angler. Black drums have mouths that point down toward the bottom. They slurp up clams and oysters and manipulate them into the back of their throat, where the pharyngeal jaws are located. These jaws are armed with huge molar-like teeth that, when driven by their powerful muscles, allow the drum to crush the shells of their prey.

Like other members of their family, this species generates low-frequency "booms" that carry vast distances through the water. The sound is generated by males during the breeding season, to attract females. The "boom" can also be used to startle potential predators. Black drums make the sound by activating muscles that line the air-filled swim bladder. These muscles contract at extraordinarily high rates of speed and are amplified by the air in the swim bladder, producing a very low-frequency sound. The frequency is low enough to travel through water, retaining walls, and earth, into homes near the water's edge, shaking walls, windows, dishes, and glasses.

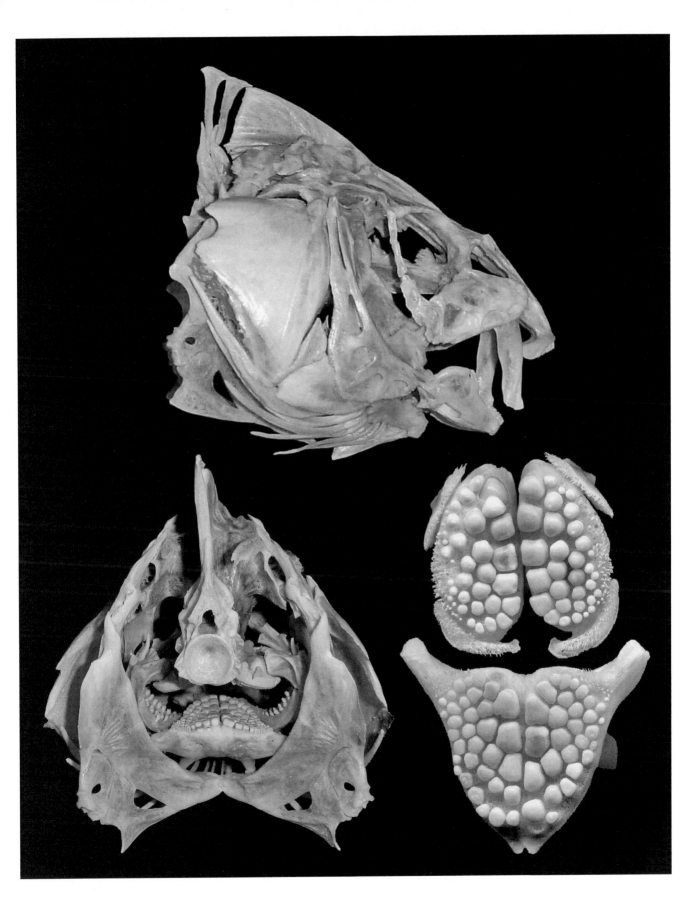

■ The spotted spoon-nose eel, *Echiophis intertinctus,* is a snake eel found on the ocean floor of the tropical western Atlantic in water up to 200 feet deep. It lives on sandy, silty bottoms, into which it can burrow to hide from threats or launch surprise attacks at prey.

Spoon-nose eels swim gracefully along the ocean floor using a snake-like pattern of waves travelling down their body, which can grow to 6 feet long. When threatened, they come to a stop and reverse into the sand or silt, tail first, until the entire body and head are completely covered. They assume this position to feed as well, striking at fish, shrimps, and squids that swim too close. Their large gape and jaws lined with sharp, needle-like teeth easily secure soft-bodied, struggling prey before it is swallowed whole.

■

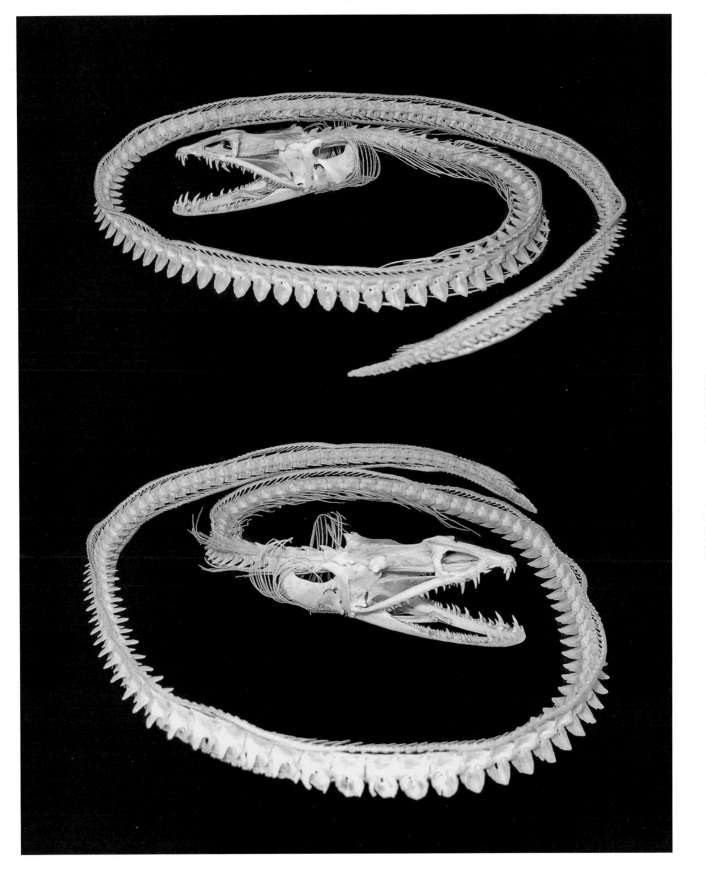

A relatively rare species of pira-nha, *Pygopristis denticulata*, or the lobe-toothed piranha, is found in the Orinoco and Amazon Rivers and some of their tributaries. It tends to inhabit clearer, calmer waters than many of its relatives and is much less prone to the marauding-school mentality of their cousins.

Lobe-toothed piranhas are opportunistic feeders. They eat fruits, insects, small fishes, amphibians, and fish eggs. Their teeth differ significantly from other piranhas' in that they are pentacuspid (having five cusps). Other piranhas have tricuspid teeth, with the middle cusp being much larger than the others, giving it a triangular appearance.

Lobe-toothed piranhas can still inflict significant damage to soft tissue; their teeth are quite sharp and are backed by capable jaw muscles and short, robust jaws that transfer energy to the teeth very efficiently.

The bridled triggerfish, *Suf-flamen fraenatum*, gets its common name from the bar-like pattern stretching back from the corners of the mouth that resembles the bridle a horse might wear. It is found from eastern Africa, through Micronesia, to the Hawaiian Islands.

Bridled triggers live near coral reefs, silty habitats, and coastal lagoons in water up to 700 feet deep, though they are usually found in much shallower water. They feed on a diversity of prey, including urchins, clams, snails, mussels, shrimps, crabs, worms, fishes, and detritus (particles of dead organisms and fecal matter). Their strong jaws and thick teeth can handle just about any type of prey they choose to attack.

The three spines on their back give triggerfishes their name. When a triggerfish swims into a crevice to hide from a threat, the first spine can be erected as an anchor. Next, the second spine locks underneath the first, securing it; this makes the first spine so stable that it will not give way unless it breaks. However, a ligament runs from the top of the third spine to the base of the second; pulling on the third spine disengages the second, and the first spine will drop like the hammer on a gun, producing the name *triggerfish*.

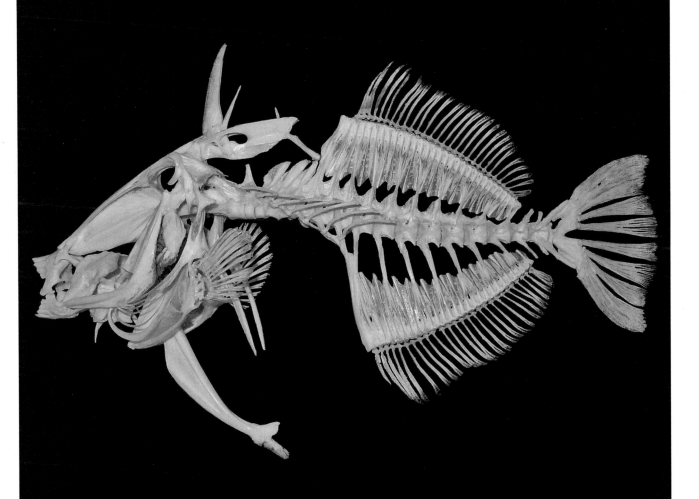

Male veiled chameleons, *Chamaeleo calyptratus*, have a huge dorsal crest on their skull that can stand almost twice the height of the rest of the skull. It is hypothesized to function as an inverted funnel that channels water toward the corners of the mouth, though sexual selection by females is also likely to be a driving factor for the evolution of such a large structure. Females have far less impressive casques, a fact that supports the sexual selection hypothesis.

Growing up to 20 inches long, male veileds are highly territorial, displaying vibrant colors and behaviors to rivals, often resulting in pinching, biting, and pushing each other until a victor is crowned. Veiled chameleons occur in a broad range of habitats, from forests to shrubs to gardens, and are native to the southwestern Arabian Peninsula, from Saudi Arabia to Yemen.

They, like other chameleons, are a hodge-podge of anatomical novelties. Their eyes can work independently, looking in two different directions at once. Their feet have two digits on one side and three on the other, making them perfectly adept at gripping branches. Their prehensile tails function like a fifth leg. Their tongues are ballistic missiles that are thrust at their food and stick by "grabbing" prey with the tip of the tongue, while the middle of the tongue creates suction to hold their meal, made up of insects, flowers, fruits, and leaves.

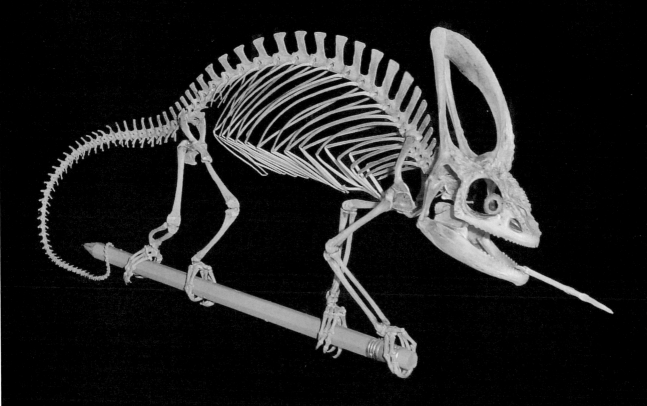
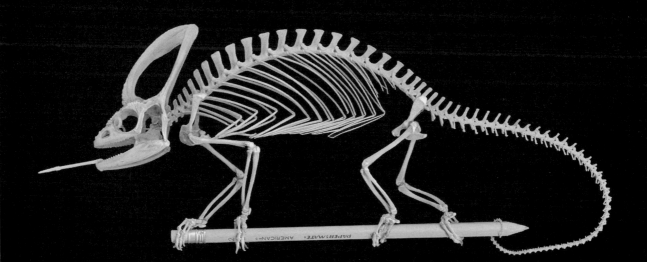

■ The muskellunge, *Esox masqui-nongy*, better known as the musky, is the largest member of the pike family, Esocidae. The species name comes from the Ojibwa word *maashkinoozhe*, which means "ugly pike."

Muskies get up to 6 feet long and can weigh nearly 70 pounds. They are found in lakes and rivers from Canada to the Tennessee River valley. Tremendous ambush predators, they hunt by hiding in vegetation and launching surprise attacks at unsuspecting prey that swims nearby. Muskellunge have enormous mouths filled with many sharp teeth, which they use to impale prey that is overtaken during ultra-high-speed attacks.

They are indiscriminant feeders on fish, crawdads, frogs, duck-lings, snakes, muskrats, and water birds. Anglers pursue them with some of the largest baits designed by lure companies, and it is not uncommon to catch a musky by accident—they often bite fish hooked by a fisherman and are then netted at the boat because they are unwilling to let go of their free meal.

■

The Bismarck ringed python, *Bothrochilus boa*, is found only near the Bismarck Archipelago off the east coast of Papua New Guinea. The only member of this genus, it inhabits rainforests and spends many of the daylight hours under piles of leaves, coconut husks, and fallen timber.

A nocturnal predator, ringed pythons spend their nights in pursuit of prey. They frequent the burrows and tunnels of fossorial (burrowing) mammals in hopes of meeting them head on or trapping them in a dead end. When prey is located, the python strikes, constricts, stopping the blood flow of their prey, and then swallows it whole. The Bismarck ringed python is one constrictor species that rarely swallows anything greater in diameter than its own body, because of its tendency to pursue prey inside restrictive, earthen tunnels.

■ The biggest piranha species alive today, the San Francisco piranha, *Pygocentrus piraya*, grows up to 20 inches long and can weigh as much as 7 pounds. Most of that weight is in the robust skull, which is loaded with large, serrated teeth and powerful jaw muscles.

This species is found throughout the Sao Francisco River drainage in Brazil. Piraya piranhas, as they're also known, spend much of their time in deeper parts of rivers, where water flow is quicker than it is near the shorelines. They tend to hide behind rocks and hunt the eddies that sometimes swirl smaller fish into their kill zone. Less social than many of the smaller piranha species, they are usually found alone.

San Francisco piranhas are opportunistic feeders and are willing to eat plant matter as well as fish. They have tricuspid teeth with serrated, razor-sharp edges that work well for slicing both animal and plant material. Their huge underbite and half-inch teeth can power through even the most fortified shell or carapace.

■ The yellow-footed tortoise, *Chelonoidis denticulata*, also known as the Brazilian giant tortoise, is the fifth- or sixth-largest species of tortoise in the world, depending on your source. It can grow to over 30 inches in length and weigh more than 100 pounds, although smaller individuals are much more common.

Yellow-footed tortoises live in grasslands, forests, and forest edges where they eat grasses, leaves, flowers, fruits, slow-moving insects, snails, worms, and sometimes carrion and bones. The serrated edges of their powerful jaws easily cut through the toughest vegetation and thickest fruit skins. Their heavy shells, thick, scaly hide, and mighty jaws serve as great protection against predators such as jaguars, coyotes, and feral dogs.

These tortoises are favorites among keepers and have been overcollected from the wild. They are currently listed as threatened with extinction in the wild because of overharvest for pets or as food locally and significant habitat deforestation and degradation.

A small, freshwater species, the Indian flapshell turtle, *Lissemys punctata*, is found in Bangladesh, India, Myanmar, Nepal, and Pakistan. It lives in rivers, streams, reservoirs, marshes, ponds, lakes, rice fields, and metropolitan canals. Highly adaptable omnivores, they eat frogs, fish, shrimp, snails, insects, flowers, fruits, seeds, and vegetation.

Interestingly, these turtles have a greatly reduced shell, in terms of both carapace (top) and plastron (bottom), because of their fully aquatic lifestyle. This is not uncommon in aquatic turtles, because the struggles of swimming with a top-heavy shell make life too difficult. As such, evolution has selected for an overall reduction in the shell to help with buoyancy and swimming efficiency.

This species is targeted for its meat and for unsubstantiated medicinal uses by certain cultures. For example, in India the shell is used as an alleged remedy for tuberculosis. In China, the shell is burnt and ground with oil to produce a conjectural medicine for certain skin diseases. However, there is no scientific evidence to support either of these claims, and the species may be in trouble due to overharvesting, pollution, dams, and development.

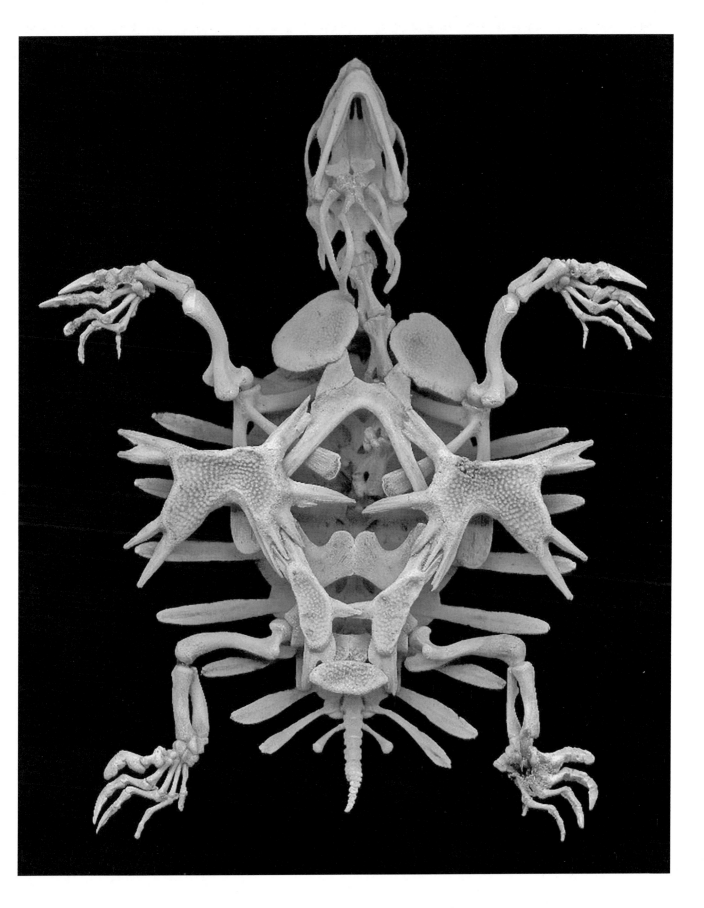

Morelia amethistina, the scrub python, is a decidedly large constrictor found in Australia, Indonesia, and Papua New Guinea. It inhabits rainforests and scrublands and is equally comfortable in suburbia, where it finds prey living among humans. Exceeding 20 feet in length, with some reports of nearly 30-footers, they are the largest snakes found in their home range.

Scrub pythons eat birds, bats, rats, opossums, and even wallabies. They commonly reside near water sources and lie waiting for prey to take a drink. Animals are grabbed, tightly squeezed to stop blood flow, killed, and swallowed whole, head first.

Like other snakes, which all lack limbs, they have the difficult task of swallowing prey usually bigger than their own head, without the aid of limbs to shove food down their throat. To swallow, they have evolved a sophisticated set of jaws, where the two rows of teeth on the roof of the mouth lie on bones that can slide front to back. When they start to eat their prey whole, these bones effectively "walk" the skull over the top of their meal, while the lower jaws do the same from below.

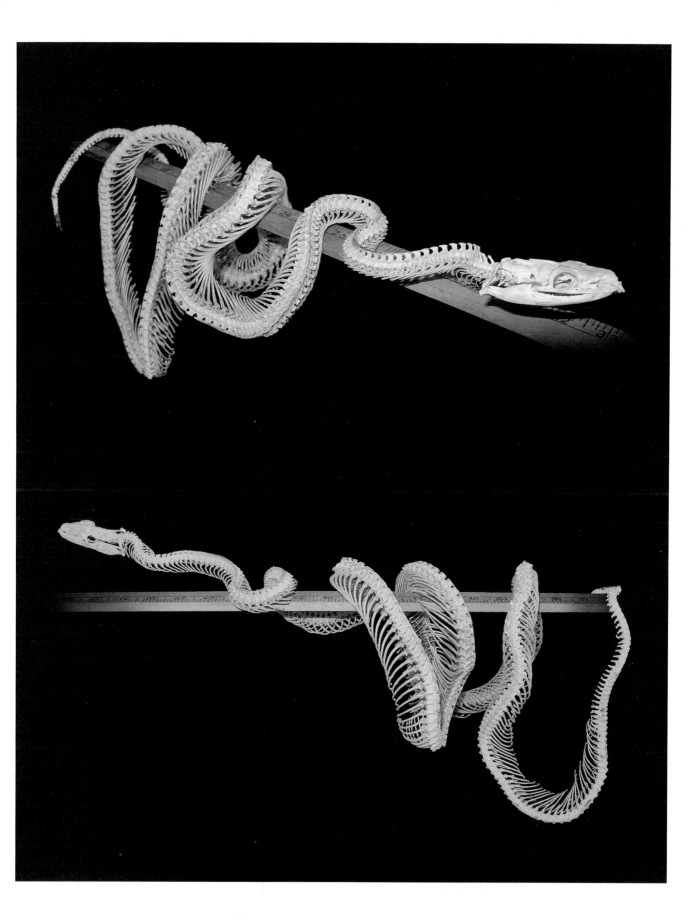

■ Shaw's cowfish, *Aracana aurita*, is an interesting-looking fish from the southern half of Australia. It inhabits coral reefs, rocky outcroppings, and seagrass beds in shallow bays and can occur in water up to 500 feet deep but is most prevalent in water less than 50 feet deep.

Like other cowfishes, the Shaw's cowfish is encased in a rigid body-covering called a cuirass. With half the cuirass removed, one can see how over-sized the pectoral bones have become to support the external shell. This shell affords them excellent protection from predators, especially since Shaw's cowfish also have multiple heavy spines protruding from their carapace. The shell, however, makes them slow swimmers with limited maneuverability—another superb example of the trade-offs that occur in nature—fast and agile or slow and fortified.

Shaw's cowfish eat a diversity of bottom-dwelling prey and have evolved a sophisticated behavior to help them locate their food. They blow jets of water at the sediment, revealing the organisms that live within it, and then pluck them out with their clipper-like teeth.

The reef lizardfish, *Synodus variegatus*, is found throughout the Indo-Pacific, Hawaii, and Red Sea on reefs and sandy bottoms in water up to 130 feet deep. It spends all of its time on the sea floor hiding from threatening animals as well as from its prey.

Reef lizardfish only get up to 9 inches long, but they eat like a much larger fish. This is because their head, which resembles that of a lizard (hence their common name), is the largest part of their body. If they can get their mouth over their food, then they will

swallow it whole. Lizardfish are voracious predators of smaller fishes, shrimps, and cephalopods and bring a mouth full of hardware to the table. Their jaws and tongue are lined with hundreds of pointed, needle-like teeth useful for grabbing struggling prey.

Lepomis macrochirus, the bluegill sunfish, is probably the most popular sport fish in North America and is certainly the species for which everyone learned to bait a hook and tie on a bobber. They are endemic to the eastern half of North America, from Florida to southern Canada, but have been transplanted all over the world.

Bluegills are the quintessential example of a suction-feeding fish. They eat insects and grass shrimp that they find clinging to vegetation and plankton that they find in the water column. Both prey require fast strikes to generate water flow into the mouth, which drags the prey in with it. As such, bluegills have large amounts of back muscle, which yank the skull upward quickly and induce the mouth to open and the mouth cavity to expand. They also have a small gape, which focuses the suction pressure directly at their prey. In fact, this species has the greatest suction pressure measured in any bony fish to date. Bluegills are common food for many other species of predatory fishes, including basses, catfishes, pikes, musky, walleye, and even crappie, when the bluegills are very small.

The African rock python, *Python sebae*, is one of the "big five" constrictors and is the longest snake in Africa. Reports of snakes greater than 25 feet long abound, but they are typically less than 20 feet long and weigh less than 200 pounds. Rocks, as they're known, are extremely thick-bodied snakes with wide midsections.

Hailing from sub-Saharan Africa, African rock pythons inhabit forests, savannas, grasslands, semi-deserts, rocky crags, and open fields, essentially anywhere they can find a meal worth grabbing. Oftentimes, rock pythons will hunt near swamps, lakes, streams, and rivers, knowing that the arid landscape means animals will have to seek out water for a drink.

Rock pythons feed on warthogs, monkeys, monitor lizards, and juvenile crocodiles in undisturbed habitats and rodents, chickens, dogs, and goats in areas where humans have encroached on their territory. The juvenile python pictured here caught a Saharan spiny-tailed lizard, *Uromastyx geyri*, and is in the process of swallowing it.

Balistes vetula, or queen triggerfish, inhabit reefs, worm-rock structures, and rocky ledges in water usually less than 150 feet deep. They occur along both shores of the Atlantic Ocean from as far north as Canada to as far south as Angola, Africa.

Queen triggerfish have very strong jaws lined with many incisor-like teeth. The species only gets up to 12 pounds and is rarely longer than 20 inches, but they have the jaws, teeth, and bite muscles of something much larger. They also have the ability to "spit" jets of water at their prey; they use this behavior to roll prey such as urchins and sand dollars over to get at the soft underbelly, which they tear into with their well-equipped mouth.

Triggerfishes get their name from the three spines on their back. The first can be erected as an anchor when the fish swims into a crevice to hide. The second spine locks underneath the first, securing it in place. At this point, the spine is so secure that it will break before it gives way. However, there is a ligament that runs from the top of the third spine to the base of the second spine. If you pull on the third spine, you disengage the second spine, and the first spine drops like the hammer on a gun—hence the name *triggerfish*.

■ The green anaconda, *Eunectes murinus*, holds the title as the heaviest snake on the planet, weighing in at over 200 pounds, with unsubstantiated reports of snakes over 400 pounds occurring in the wild. Anacondas are the second-longest snakes alive today, exceeded only by the reticulated python, *Python reticulatus*. They can grow in excess of 20 feet, again with unsubstantiated reports of giant anacondas over 30 feet long encountered in the wild.

Inhabiting swamps, marshes, and streams, anacondas are highly accomplished swimmers; they submerge just under the surface and lie waiting for their prey. When an unsuspecting capybara, bird, fish, or caiman (like this baby black caiman, *Melanosuchus niger*) gets too close, the anaconda lashes out with an explosive strike to snatch the prey with its mouth and then subdues it with strong coils around the body. Many people think one dies of asphyxiation when coiled by a constrictor. In fact, death usually occurs long before you run out of air; the coils stop blood flow through your arteries and veins, which causes your heart to stop beating, and you die of a heart attack.

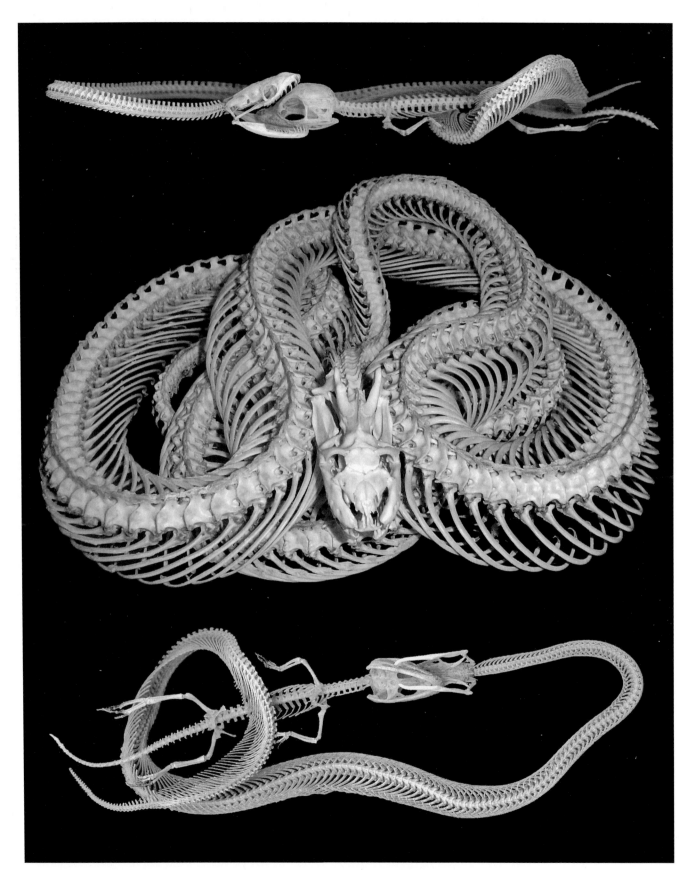

■ The white-throated monitor, or rock monitor, *Varanus albigularis*, is the second-longest lizard on the African continent at greater than 6 feet, second to the Nile monitor, *Varanus niloticus*. White-throateds are, however, the heaviest lizards in Africa, at up to 40 pounds.

They occur throughout south-western, south-central, and eastern Africa in dry habitats, such as savannas, steppes, prairies, and rocky slopes. They move very quickly and are quite agile when chasing prey, which includes large insects, rodents, lizards, snakes, and any eggs they find. Their short, blunt teeth are used to pulverize anything they can chew so the monitor can more easily swallow it, whole, down its gullet.

Rock monitors are strong animals with robust legs, a whip-like tail, sharp claws, and forceful jaws. They stand their ground against threats such as honey badgers and will posture, hiss, slap, bite, and claw at anything trying to subdue them.

■

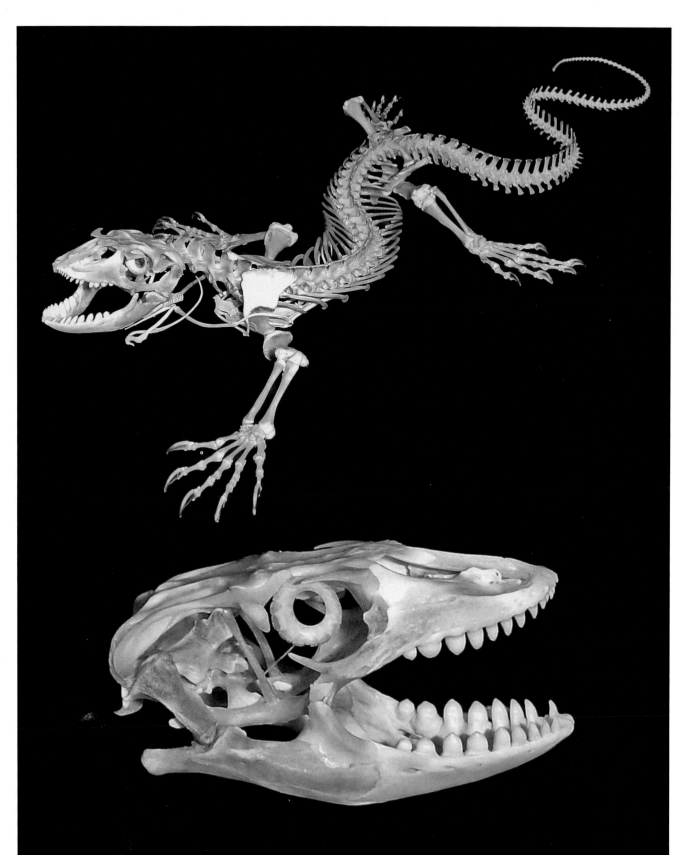

■ Zebra moray eels, *Gymnomuraena zebra*, are found throughout the Indo-Pacific on reefs and rocky bottoms in water up to 100 feet deep. They establish a lair within the reef structure and spend their days hiding inside it. At night they come out to pursue prey, a behavior different from nearly all other moray species.

Zebra morays grow up to 4.5 feet long and eat very hard prey, such as clams, urchins, and crabs. Other morays usually have long, thin jaws lined with sharp, needle-like teeth that they use to capture elusive, soft-bodied fish and shrimp. Conversely, zebra morays have short, robust jaws lined with molar-like teeth. They grab hold of something hard, roll it around in their mouth to locate a weak point, and employ mighty jaw muscles to crush the shell or exoskeleton. The soft, edible portions are then pulled back to the throat with highly mobile pharyngeal jaws in the rear of the mouth. These jaws slide forward, grab the prey, and retract to the esophagus. This zebra moray had apparently consumed only purple sea urchins for most of its life and had absorbed the urchins' pigment, as its skeleton is a deep magenta color.

Russell's vipers, *Daboia russelii*, are considered the most lethal snakes on the planet because of the number of people that die from their bites each year. Their venom is delivered through long, mobile fangs and is highly potent. The toxins cause severe pain, swelling, bleeding from the gums, paralysis of the neck muscles, and most critically, kidney failure. Even those who get treatment in enough time to survive usually experience kidney problems for life.

This species has a very wide distribution, occurring throughout the Indian subcontinent and much of Southeast Asia. They are often found around agricultural regions, where they hunt for mice, rats, shrews, squirrels, and lizards attracted to the area by ample food.

Russell's vipers grow up to 5 feet long, are relatively thick bodied, and have an extremely fast strike for their size. When threatened, they curve into a repeated S shape, raise a third of their body off the ground, hiss loudly, and strike with such force that they appear to "jump" at the threat. Their genus name means "the lurker" in Hindi, due to their tendency to hide in close proximity to humans.

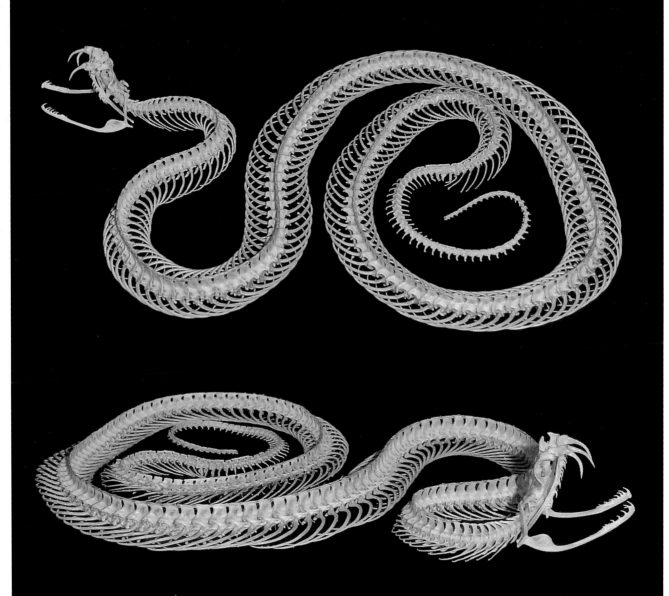

■ A *Boa constrictor*, as the name implies, kills by wrapping a death grip around its prey and constricting until blood ceases to flow. These large-bodied snakes inhabit arid regions from Mexico to Argentina, spending the least amount of time near water of all the constrictor species. They spend their days coiled in a protected location and come out to feed at night.

Boas are equally comfortable on the ground and in trees. When crawling about on the ground they feed on amphibians, lizards, and small mammals. When moving through the trees, they target perched birds or bats as they fly by in complete darkness. Boas use heat-sensitive scales near their mouth to pinpoint warm-blooded prey, including bats; they snag them right out of the air and into their coiled bodies before swallowing them whole.

There are many varieties of *Boa constrictors*, and breeders are creating new patterns every year. The market for this snake in the pet trade is significant enough that wild populations are considered threatened, even with the significant amount of captive breeding that occurs. Some subspecies that inhabit specific Caribbean islands are completely protected from the trade.

■ Smooth butterfly rays, *Gymnura micrura*, are found in the western and eastern Atlantic Ocean. They inhabit soft-bottom, marine environments and can even thrive in low-salinity estuaries and high-salinity lagoons.

Rays, like sharks and skates, completely lack bone in their skeletons. Instead, they are composed entirely of cartilage, which is much more flexible and watery than bone. The starburst pattern of filaments throughout their body is formed by structures called ceratotrichia, onto which muscles attach.

Bony fish, crabs, shrimps, and snails are the food of smooth butterfly rays. They may have more than 200 teeth in their mouth, each with a thick base and sharp tip, perfect for piercing soft prey yet strong enough to crush hard prey. Lacking a venomous spine, they are therefore preferred prey for many larger fish and mammals.

For numerous reasons, flathead catfish, or *Pylodictis olivaris*, are some of the most sought-after fish in North America. Reaching 4 feet long and 125 pounds, they get very large, which makes them challenging and fun to land for anglers. They are prized food fish for the giant fillets they yield for catfish fry-ups. And they are the main target of noodlers who use their bare hands to snatch flatheads out of their underwater lairs.

This species inhabits slow-moving rivers, backwaters, lakes, and reservoirs of the central and southeastern United States but has been transplanted as far west as California. They feed on anything that fits in their mouth, including fish, crawdads, insects, worms, snakes, frogs, mudpuppies, ducklings, turtles, and even small mammals such as muskrats. Like other catfishes, they lack scales, and their bodies are covered with taste buds for tracking down their next meal.

Contrary to popular belief, the whiskers of a catfish cannot sting. They are simply extensions of the skin to assist in localizing the source of a meal. The threat from catfishes, flatheads included, comes from the barbed spines on the front edge of their pectoral and dorsal fins. The side of each spine is lined with rear-facing barbs that do not allow the spine to be removed without significant tissue damage.

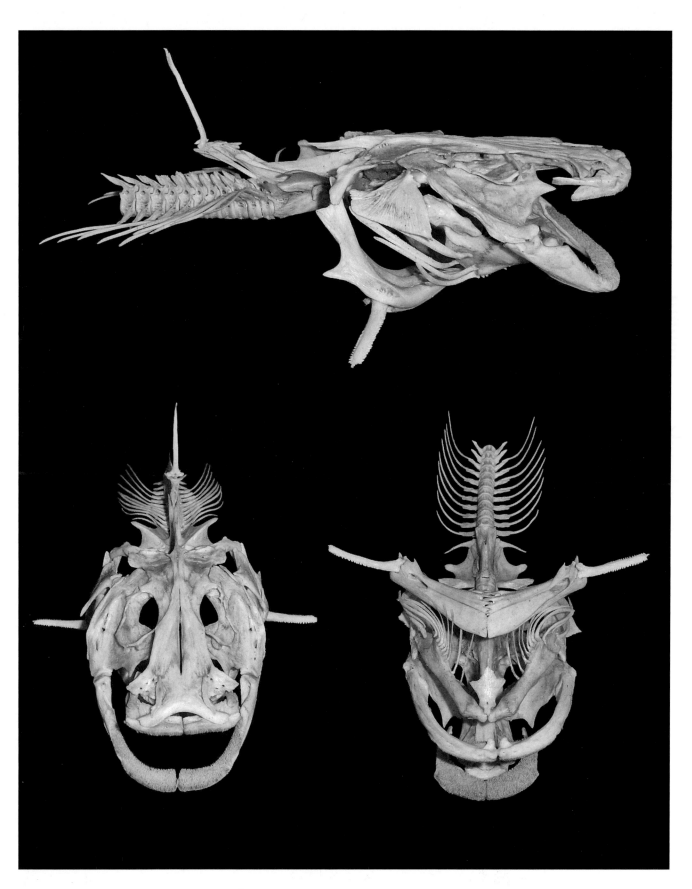

■ Black and white spitting cobras, or Indochinese spitting cobras, *Naja siamensis*, are listed as vulnerable because of their continued overharvest from the wild for use in traditional Chinese medicine, which is entirely illegitimate and driven by ignorance. They occur in lowlands throughout countries such as Cambodia, Laos, Viet Nam, and Thailand, where they are persecuted by landowners and farmers due to fear.

Spitters can deliver their venom via a bite or through their uncanny ability to target the eyes of a threat in hopes of blinding them so they can sneak away to safety. If the venom is delivered through the skin, it makes its way directly into the bloodstream and will cause muscles, including the diaphragm, to be paralyzed, so envenomated animals and humans die of asphyxiation. If delivered to the eyes, it may cause permanent blindness.

Like other cobras, *Naja siamensis* eat small mammals, frogs, toads, and other snakes. They are a critical part of managing rodent populations, especially in rice paddies and croplands.

■

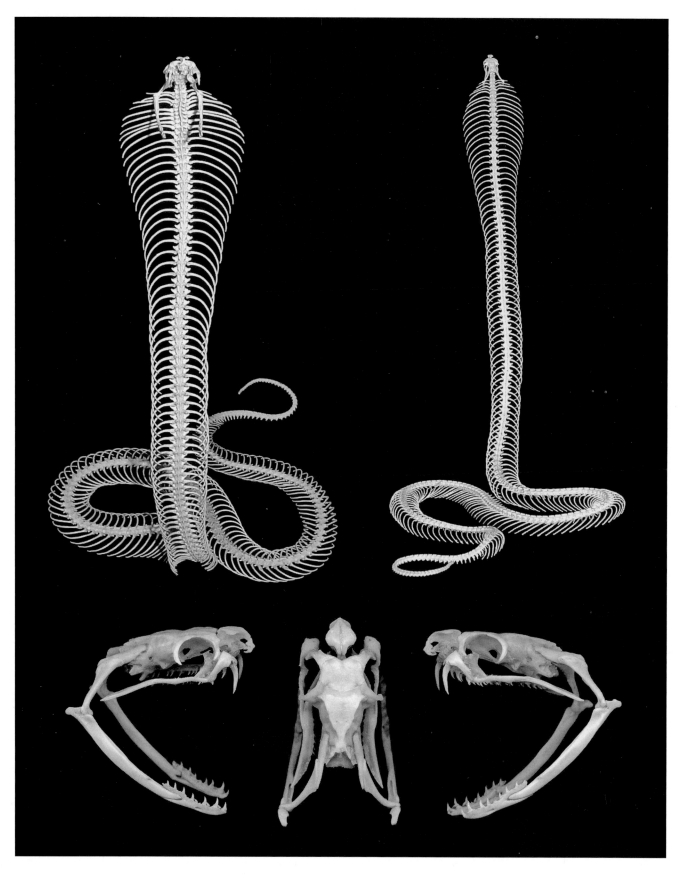

■ Aptly named, the bonefish, *Albula vulpes,* is considered the holy grail of saltwater fly-fishing. On the flats of the Florida Keys and the Bahamas, anglers wade and push-pole around, trying to find a bonefish with its nose down in the sediment searching for food. A well-placed fly gets gobbled up, and the fight of a lifetime is under way. Bonefish are active fish living in warm water, so they have to eat often and will chase whatever tempts them—hence their targeting by anglers.

Bonefish eat just about anything they can find in the sediment, including worms, shrimp, crabs, bivalves, and fish. They are able to process such a diversity of prey because of the suite of variable teeth lining their mouth.

There are small, pointed teeth on the oral jaws, which help to capture prey, and rounded molars on the tongue, roof, and sides of the mouth to crush hard stuff.

■ Some of the bluest blue in the ocean occurs in the skin of the scrawled filefish, *Aluterus scriptus*. It inhabits reefs and lagoons in brightly lit water, usually less than 75 feet deep, and is found in tropical and some subtropical locations all the way around the globe.

Large for a filefish, they grow to more than 3 feet long and are a favorite prey of many reef predators because they do not swim very fast. As such, the blue in their skin, coupled with dark spots on an olive-drab background, serve as excellent camouflage to hide from potential threats, such as mahi mahi, tunas, groupers, and barracudas.

Scrawled filefish, also known as broomtail filefish, move slowly about the reef so as to not attract the attention of predators while they look for soft prey. They eat mainly algae, sea grasses, gorgonian corals, anemones, and tunicates. Their upward-facing snout and clipper-like jaws are perfect for snipping off pieces of prey with their human-like incisors.

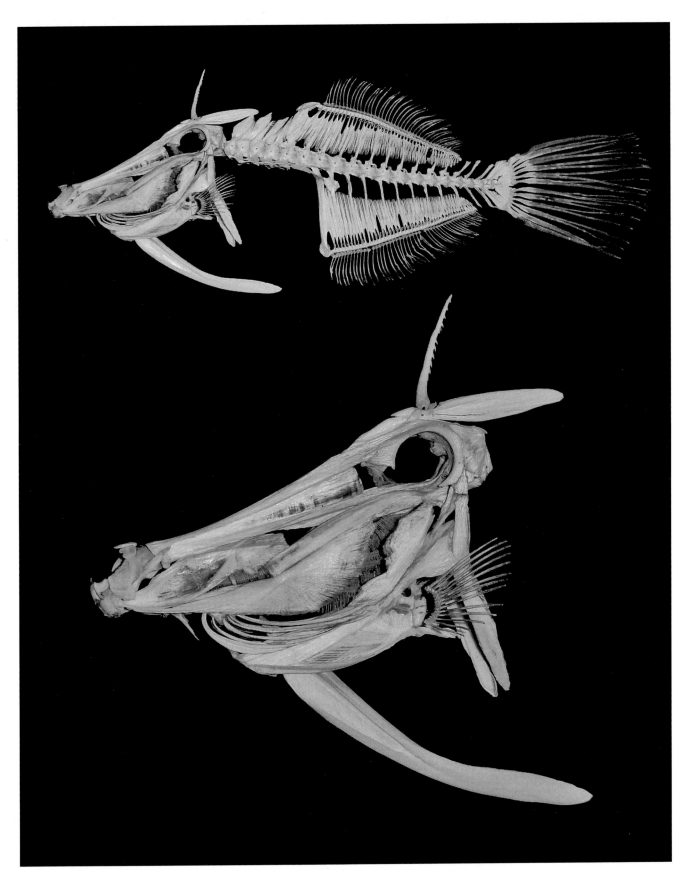

■ The bottom-feeding searobin, *Prionotus carolinus*, lives on sandy bottoms and in channels down to 600 feet deep from Nova Scotia to the Gulf of Mexico. They grow up to 16 inches long and are common prey for larger benthic species, such as flounders and some sharks.

To deter predators, searobins have evolved a specialized set of pectoral fins than are usually collapsed along their sides when at rest. If approached by a potential predator, searobins will quickly throw their pectoral fins open like a bird's wings, revealing one large false eyespot on each fin. To the approaching threat, a giant pair of eyes just opened on the sea floor, and the attacker quickly redirects to safety.

Another specialized feature of the pectoral fins of searobins is in the last three rays, which are curled and have independent muscular control. Searobins use these rays like fingers to "walk" along the bottom as if traveling on their fingertips. Here they hunt for shrimps, crabs, squids, bivalves, and smaller fish.

■

The spiny softshell turtle, *Apalone spinifera*, is a large, freshwater turtle growing to nearly 20 inches long. It occurs throughout the central and eastern United States and is found as far north as Canada and as far south as Mexico. This turtle thrives in nearly all freshwater habitats, from rivers to ponds to the Great Lakes.

Softshells, as they're known, are carnivorous and will eat anything they can grab off the river or lake bottom, including worms, crawdads, fish, and frogs. They tend to move along the substrate between rocks and around vegetation, snatching up whatever tries to flee.

As a fully aquatic turtle that must breathe air at the surface to survive, evolution has selected for a reduction in the overall size and composition of its shell. As the common name implies, the shell is not rigid, which makes it lighter. Also, compared to most turtles, much less of the carapace (top shell) and plastron (bottom shell) is bony. This feature reveals the tips of the ribs so often hidden by the large, robust carapace of terrestrial turtles.

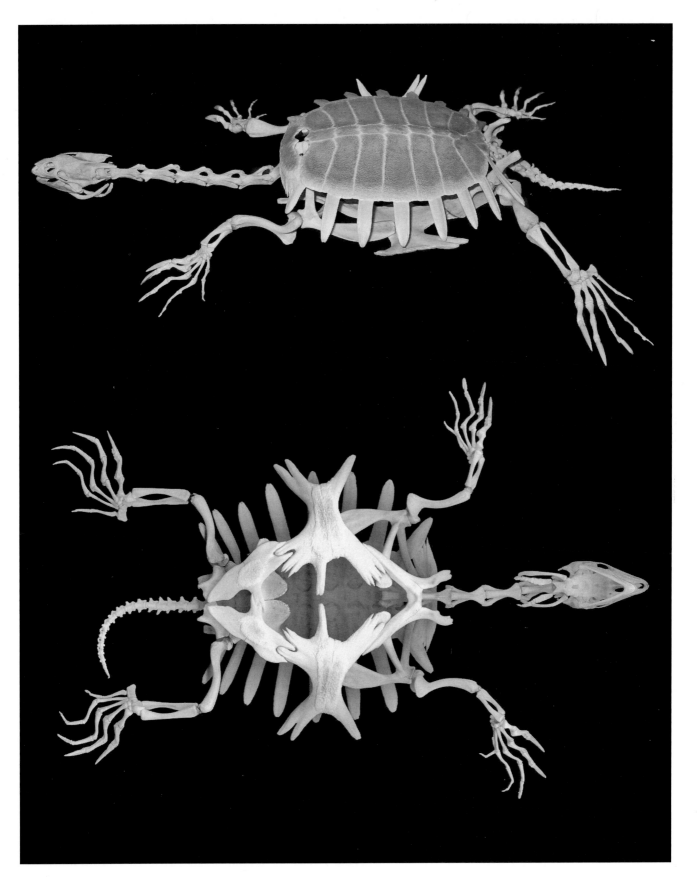

■ Lethal. No word better describes the great barracuda, *Sphyraena barracuda*, and the hardware it brings to the dinner table. These fish function as the lions of the reef, eating anything that appears weak or injured, not that they can't outswim just about any healthy fish as well. They tend to hang out near any kind of structure, because it helps them stay concealed before they launch their surprise attacks.

Barracudas are known to be predators on small- to medium-sized fish that make the mistake of swimming too far from the safety of structures. They are ram-feeders, which means they need ample room to overtake their prey as they swim through it. Cudas hit their prey at a velocity and with a force that is nearly unrivaled in the oceans. When coupled with a switchblade-sharp mouthful of teeth and a set of jaws that work like a scissor jack, they can bite through almost anything too slow to escape.

Thankfully, barracudas do not target humans as prey. The only known attacks on humans involved barracudas caught on fishing tackle or speared under water. There have even been a couple of bites on the wrist of people diving into the ocean off a dock while wearing a shiny watch. The barracuda hiding under the dock mistook the watch for a shiny fish and in less than a second, the damage is done. They are a highly inquisitive fish, as anyone who has ever dived with them knows, but they pose no real threat to humans.

■

The inland taipan, *Oxyuranus microlepidotus*, also known as the fierce snake for obvious reasons, is found on the black soil plains of central Australia. The region is nearly devoid of vegetation, so the snakes use cracks in the dry soil to move about while hiding from predators and the intense sun.

Very fast and agile snakes, inland taipans grow up to 8 feet long. While they usually flee from threats, they will gladly stand their ground if cornered or captured. Their venom is considered the most toxic of all snake venom in the world, easily capable of killing a human. The venom is usually employed to kill the booming populations of rodents in their region, but other small mammals are also on the menu. It has a highly neurotoxic component that shuts down the actions of the nervous system, while hemotoxic elements cause blood to clot within blood vessels, thereby stopping blood flow to the heart or brain.

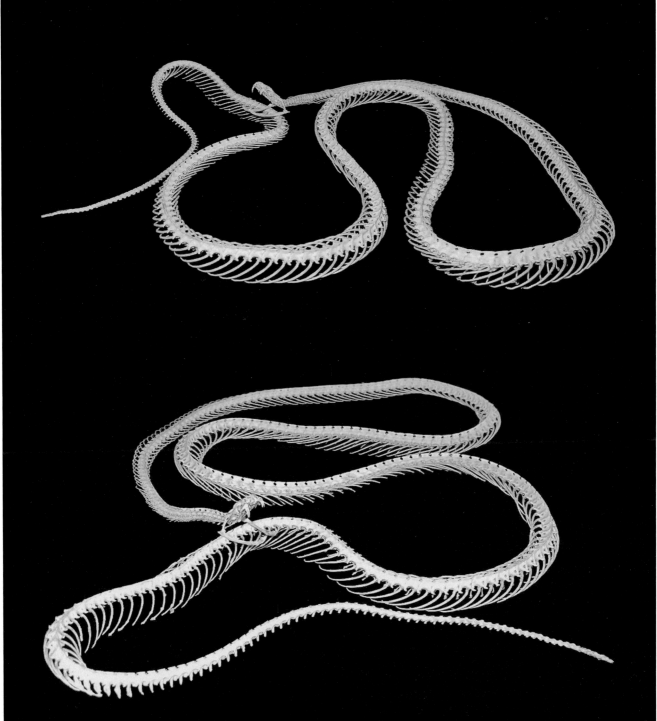

Growing up to 16 inches long, the black spot piranha, *Pygocentrus cariba*, is recognized as a potential threat to anything that enters its waters. With large, strong jaws and spade-shaped, tricuspid teeth sharp enough to slice through anything organic, it is certainly one of nature's most honed carnivores.

This species is found in the Orinoco River and its tributaries as well as the Llanos in Venezuela. Black spot piranhas get their name from the large, dark spot located just behind their gill cover. They typically occur in small schools and are known to be quite aggressive in the wild.

The popular videos of piranhas congregating under nesting birds, waiting for a clumsy fledgling to fall into the water, are most often black spot piranhas. They usually feed on other fishes and do so by swimming into a school of smaller fish and snapping at whatever they can sink their teeth into. Injured prey are often carved up and swallowed by multiple school-mates before they even stop breathing. Other prey includes snakes, frogs, capybaras, juvenile caiman, and occasionally livestock.

The purple rose queen cichlid is a cross between a Midas cichlid, *Amphilophus citrinellus*, and a red-spotted cichlid, *Vieja bifasciatum*. Cichlids, a very diverse family of fish found in South America and Africa, have the potential to hybridize more regularly than other families. Aquarists manipulate this tendency to design new variants of cichlids for sale in the industry.

Purple rose queens get their name from their vibrant pink and fuchsia colors. They grow to over 6 inches long, weigh up to half a pound, and seem to have a much less aggressive demeanor than their full-blooded Midas cousins. They tolerate fellow tank residents well and do not "rearrange" their homes like both their parent species do in captivity.

Like their parent species, rose queens are omnivores and will eat just about anything offered to them, from goldfish to pieces of squash. Their upper jaw can protrude far off their face, which helps with capturing prey of diverse shapes, sizes, and elusiveness.

A perfectly built predator, *Astroscopus y-graecum*, the southern stargazer, is found on the ocean floor of the subtropical and tropical western Atlantic in water up to 200 feet deep. It lives on sand, silt, and rubble bottoms into which it can burrow to hide from threats or launch surprise attacks at prey. It burrows by wiggling its body, flapping its fins, and inhaling and exhaling water to create a fluidized bed into which it can submerge.

Stargazers' eyes, nostrils, and mouth are perfectly positioned to reside just above the bottom while the rest of the body is buried out of sight. They lie in wait for fish, shrimp, squid, or octopi to swim or crawl nearby and then strike with absolute ferocity by exploding out of the sand and engulfing their prey. Stargazers also possess a novel means by which they generate an electrical current with special organs located in modified eye muscles. To deter predators such as rays and sharks, stargazers can expel this electrical current into the water above them.

The weedy scorpionfish, *Rhino-pias frondosa*, is a relatively small 10-inch fish found on the shallow reefs of the Indian and western Pacific Oceans. It spends its time hiding on the reef with superb camouflage that resembles a rock or other reef structure. Weedy scorpions possess pigments that match the reefs on which they reside, and they have fleshy protuberances that break up their outline and contribute to their common name.

They also tend to slowly scuttle along the reef, using their fins to move about while they sway from side to side, resembling a piece of algae. This unassuming behavior allows them to creep up on prey, which they then explosively inhale. Their large mouth and expandable stomach allow them to eat prey almost as big as themselves, and since they sometimes go weeks between meals, this adaptation serves them well. They can lie waiting for days, sitting motionless while resembling the reef, waiting for some unsuspecting small fish or shrimp to swim nearby.

For protection, they have venomous dorsal spines capable of delivering a toxic assortment of compounds. This cocktail brings excruciating pain and swelling to those that accidentally get stung and to those predators that try to eat weedy scorpionfish.

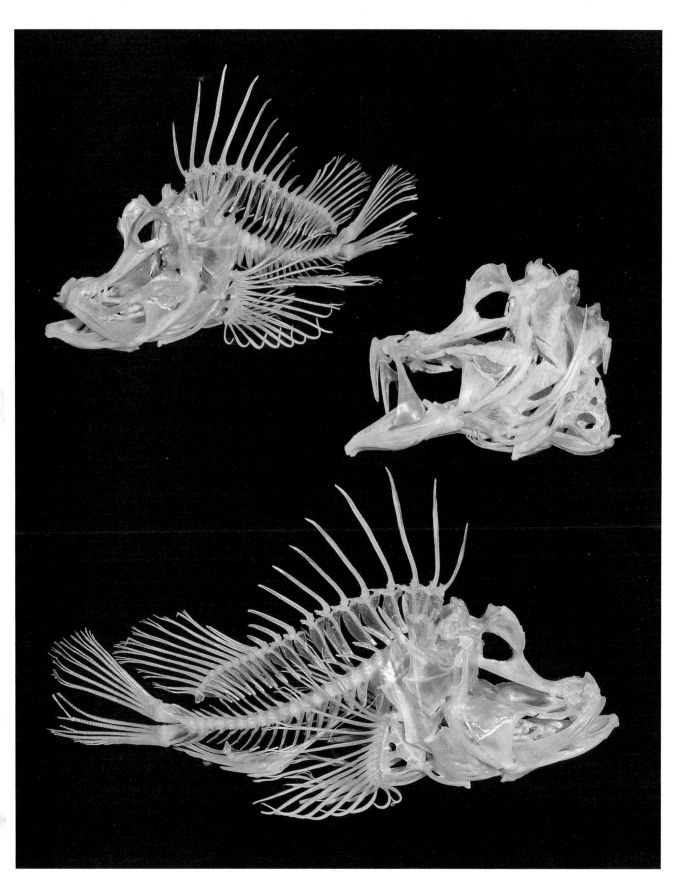

A stocky reptile, the Egyptian spiny-tailed lizard, *Uromastyx aegyptia*, is found in many Middle Eastern countries, including Egypt, Iran, Iraq, Saudi Arabia, Jordan, and others. It inhabits rocky, gravelly, stony areas on flat to hilly slopes, where it digs burrows in the ground and spends most of the day foraging on low-growing vegetation close to its den.

Spiny-tailed lizards get their name from the pointed scales that ring their tail. The tips of these scales are strong and sharp and can be used to lash out at a threat when the tail is swung like a whip. They are also known to reside head first in their burrows, a position that leaves the pointy, hard tail blocking predators that may try to snatch them out while the lizard digs in its long, sharp claws as anchors.

Egyptian spiny-tailed lizards can grow up to 3 feet long and have powerful jaws lined with thick, grinding teeth that are driven by big muscles. Their sometimes ferocious bite is needed to defend against predators, including the ill-informed people who overharvest them for use in traditional medicines.

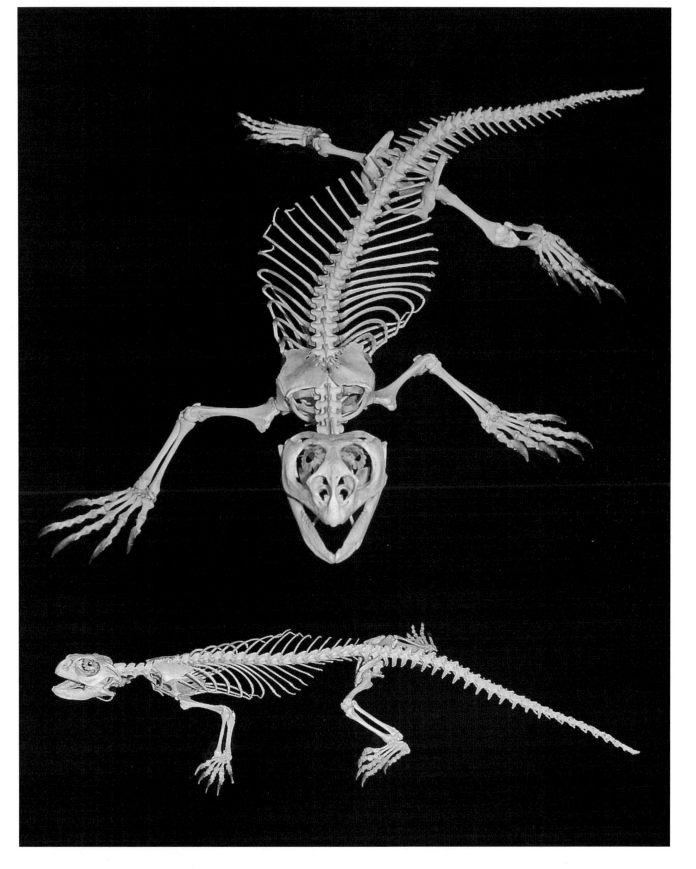

■ *Bothrops atrox*, also known as the fer de lance (loosely, French for "spearhead"), is a venomous pit viper found in the tropical lowlands of northern South America, including coffee and banana plantations, and accounts for more human envenomations than any other species where it occurs. They are generally nocturnal but may hunt during the day if the need arises. They are also generally terrestrial, though they are accomplished swimmers and will even venture into trees, especially as juveniles.

Fer de lances acclimate well to disturbance, such as the clearing of rainforest for human habitation. As such, they frequent building edges and garbage piles in search of small mammals, often leading to human encounters. Left alone, this snake will flee to safety; when threatened, they are aggressive and unpredictable.

■

■ *Gomphosus varius*, or the bird wrasse, is a member of one of the most diverse families of fishes on the planet, the Labridae. It inhabits coral reefs and lagoons over a broad range of the Indo-Pacific in water up to 100 feet deep.

Like many members of the labrid family, bird wrasses are sequential hermaphrodites, which means they switch sex during their lifetime. They begin life as female and become part of a harem protected by a large male. If the male is removed by a predator or dies of old age, the alpha female quickly becomes the new harem-protecting male by changing sex.

Bird wrasses have a specialized set of jaws, much different from most other wrasses'. In fact, their genus name is derived from the Greek word *gomphos*, meaning "nail." This could refer to their long, pointed jaws or to the fact that the jaws are tipped with pointed canine teeth. They use these canines to grab bottom-dwelling crustaceans, fishes, brittle stars, and mollusks.

■ The most common pet turtle in the United States is the red-eared slider, *Trachemys scripta elegans*. They are native to the Mississippi River drainage as well as states as far west as Colorado. Common in the southeastern United States, they have also been introduced around the world and are considered one of the top 100 invasive species on the planet.

Red-eared sliders grow to 16 inches long and can live for decades under the right conditions. They are semi-aquatic turtles and are often seen in large groups, sunning themselves on logs, rocks, or banks to warm their bodies. As juveniles, they are more carnivorous than adults, eating insects, worms, small fish, and carrion. Adults consume more vegetation than meat but are certainly opportunistic feeders.

Like other turtles, tortoises, and terrapins, they have a keratinized beak lining their jaw bones. This beak is built from the same tissue as claws and horns and, in red-eared sliders, is very sharp for cutting through prey or defending themselves against threats such as raccoons.

■ One of the smallest rattle-snakes, the eastern massasauga, *Sistrurus catenatus*, can be found further north than any other rattlesnake species. They are distributed across the central United States, are found as far south as northern Mexico, and occur as far north as Ontario, Canada. Wetlands and bordering woodlands are their main habitations. Probably due to their occurrence in river flood plains, the Chippewa Indians named them "massasauga," meaning "great river mouth."

Massasauga rattlesnakes rarely exceed 2 feet in length and spend their days in search of frogs, lizards, small mammals, and centipedes. They frequent mammal burrows in search of voles and shrews, their preferred food.

Although human deaths are very rare, massasaugas have potent venom. The venom is delivered via hypodermic-like, retractable fangs that quickly envenomate their prey, which is then swiftly released. Containing digestive enzymes, among other things, the venom kills and partially digests their prey, making them easier to swallow once found again. As the prey wanders off to die, it is located through the use of the snake's forked tongue, which helps it determine the direction of its meal's scent trail.

Gopher snakes, *Pituophis cat-enifer*, also known as bull snakes, are one of the most widespread snakes in North America, being found from the Atlantic to Pacific coasts and from Canada to Mexico. They live in every habitat imaginable, including forests, grasslands, deserts, and farmlands, and survive at elevations up to 9,000 feet. This species can get up to 9 feet long and is known to be an accomplished climber.

Gopher snakes are willing to stand their ground against potential threats. They will puff up (by inhaling extra air and spreading their ribs), hiss loudly, raise their head, and shake their tails under the leaves to resemble the rattle of a rattlesnake. Their skin is even patterned like a rattlesnake's, which often leads to mistaken identifications by people.

As a member of the colubrid family, they kill by constricting their prey, which consists mainly of mammals. However, these accomplished swimmers will also pursue frogs in ponds, and as excellent climbers, they will eat birds and their eggs.

Their common name, sailfin tang, comes from their extremely large dorsal and anal fins, which increase their profile by 100%. Their scientific name, *Zebrasoma velifer*, is derived from their stripes, which resemble the African zebra, plus the Greek word *soma*, meaning "body."

Sailfin tangs are found on tropical coral reefs and lagoons in the western Indian Ocean in water usually less than 100 feet deep. They grow up to 16 inches long and are found alone or, especially during the breeding season, in pairs. Sailfin tangs eat leafy algae and have more pronounced pharyngeal teeth than other tangs and surgeonfishes. They use these specialized teeth to grind the large pieces of vegetation that make up their diet.

A sailfin's large dorsal and anal fins act like a kite when the fish needs to turn sharply to avoid a predator or defend its home territory against a rival. These fins can be collapsed in order to make the fish more hydrodynamic when they try to swim quickly, but they can be swiftly erected when it's time to make a hairpin turn. The single caudal (tail) spine can be used to slash anything perceived as a threat or a competitor.

The largest species of hognose snake in the world, the Malagasy giant hognose, or *Leioheterodon madagascariensis*, reaches 6 feet in length and is a very thick-bodied snake. It is native only to the island of Madagascar and inhabits sandy scrublands and loamy forests, where it can easily burrow into the soil to pursue prey or bury its eggs. The up-turned "hognose" for which the group is named is a feature that evolved to accommodate their burrowing habits.

Giant hognoses eat small birds, rodents, amphibians, lizards, and lizard and snake eggs that they find buried in the soil. Hognoses are known to be slightly venomous and inflict their venom using fangs positioned much farther back on the jaws than most other venomous snakes. Because of this, they can often be seen gnawing on prey in order to get the jaws into position to inject the venom. The mild venom serves to relax their prey, making them easier for the snake to swallow.

Red gurnards, *Chelidonichthyes cuculus*, occur in water up to 1,300 feet deep throughout the temperate eastern Atlantic, including the Mediterranean Sea. They are bottom-dwelling fish, patrolling the ocean floor in sandy, rocky, and gravelly areas in search of prey. Due to their habit of never leaving the sea floor, evolution has selected for an armored dorsal surface while eliminating armor or spines from the underside. Gurnards have thick, robust plates over their skulls and numerous long, dorsal spines for protection from above.

Red gurnard pectoral fins have three finger-like rays at the bottom that they use to "walk" along the seafloor as if they were on their tip-toes. They scuttle along and snatch up any small crustacean, fish, worm, or other soft-bodied invertebrate they find.

Reaching up to 16 inches long, red gurnards are a favorite prey of many Atlantic predators. Along with their protective plates and spines, they have a secret weapon. When threatened, they can throw their large pectoral fins out quickly to reveal two large, false eyespots, making the predator think that something even bigger just opened its eyes below them.

The yellow-bellied watersnake, *Nerodia erythrogaster flavigaster*, is a commonly encountered constrictor throughout its home range. It lives in swamps, ponds, shallow lakes, and river back-waters and is a highly capable swimmer and climber.

Yellow-bellied watersnakes eat mainly frogs but will also glad-ly sample crayfish, toads, fish, tadpoles, salamanders, and very small mammals. They grow up to 6 feet long and are considered pugnacious when cornered, as one should expect of any animal fearing for its life.

Accomplished swimmers capable of holding their breath for longer than 30 minutes, they often dive to the bottom and rummage around logs or rocks, where they surprise underwater prey with fast, toothy strikes. They are even reported to hang from overlying trees with their head underwater to capture small fish that swim close. Prey is usually carried to a bank, where the snake can kill it, reorient it to a head-first position, and swallow it whole using sophisticated, ex-pandable jaws and stretchy skin.

Commonly encountered, the corn snake, *Elaphe guttata*, has vibrant red, yellow, and orange skin. It occurs throughout the southeastern United States but is most common in Florida and bordering states. People often encounter this species because it's diurnal (active during the daytime) and because it readily visits manmade structures in search of rodent prey. It also lives in places where humans build homes, such as woody groves, forests, prairies, and hillsides.

Corn snakes often hunt prey by traveling through their underground networks of tunnels and will eat mice, voles, shrews, rats, and lizards. When they happen upon something edible, they quickly grab it with their mouth and wrap a couple of coils around its body to kill it. Meals are swallowed whole and can be bigger in girth than the snake's head and body. Sophisticated jaws with multiple specialized joints and stretchy skin make this possible.

The raccoon butterflyfish, *Chaetodon lunula*, is an 8-inch butterflyfish from the shallow waters of the Indo-Pacific. Its showy color pattern—a background of yellow with black stripes and blotches and a large white band behind the eye—makes it easy to spot on the reef. There is also a false eyespot, located near the base of the tail, which tricks predators into attacking the wrong end of the fish. Their dorsal, anal, and pelvic fins are full of thick, sharp spines that make them difficult to swallow and, with luck, cause predators to release them.

Raccoon butterflyfish use their pointed snout to capture and feed on sea slugs, tubeworm tentacles, small anemones, coral polyps, and algae. They inhabit lagoons, reef slopes, ledges, and rocky areas with lots of structure, on which they find their prey. A common fish within the aquarium trade, they tend to get along well with most other species in captivity.

A colossal flatfish, the Pacific halibut, *Hippoglossus stenolepis*, is found in northern Pacific waters at depths exceeding 1,000 feet. This species can grow to over 500 pounds and be longer than 8 feet. It eats whatever it can fit in its mouth, including octopi, crabs, and many other species of fish.

Like other flatfishes, they begin life like any other larval fish—bilaterally symmetrical, with one eye on each side of the head. However, very early in their lives one eye migrates over the top of the skull and takes up residence on the other side. Simultaneously, the fish turns to create an eyeless side that lies on the ocean floor and an eyed side that watches above. Evolution drove this feature in flatfishes because it renders their three-dimensional world a two-dimensional one; this adaptation results in no predatory threats from below and great vision above.

Halibut and other flatfishes also have quite different coloration from blind to eyed side. The blind side is stark white and entirely pigmentless, because these fishes nearly always lie on the bottom, and pigment is energetically costly to produce and maintain. Meanwhile, the eyed side is perfectly cryptic for hiding from threats from above, such as killer whales and salmon sharks, and for creeping up on prey.

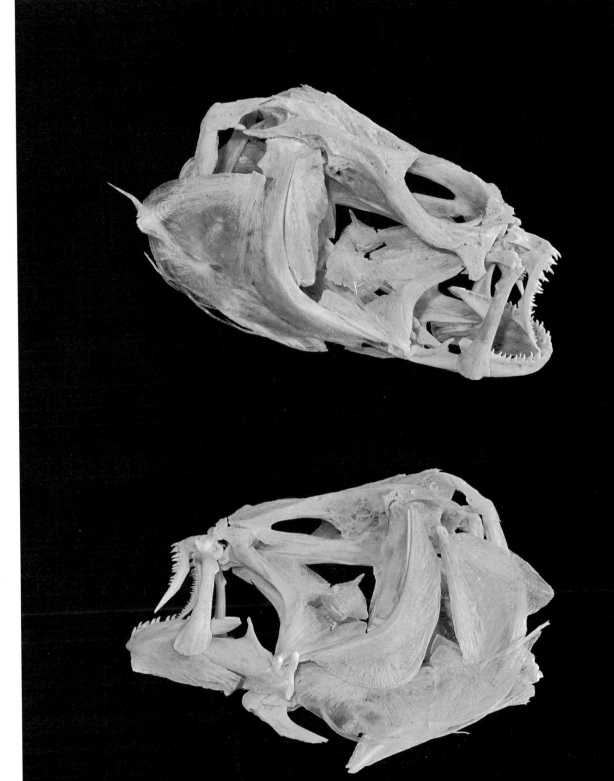

■ The Saint Lucia lancehead, *Bothrops caribbeus*, is an extremely rare species of venomous snake endemic to only one island in the West Indies. Though it is one of the smallest lanceheads, often less than 4 feet long, it has some of the longest fangs in the world and uses them to deliver a poorly understood toxin. One man experienced a stroke after being bitten by this species, which implies that the venom has some effect on the cerebral blood vessels, a most uncommon target of snake venom. Other effects include tissue hemorrhage due to the inhibition of clotting proteins, as well as platelet clumping, which clogs blood vessels.

St. Lucia lanceheads use their long fangs to administer venom into prey such as small mammals, birds, lizards, frogs, and other snakes. Due to their restricted home range and persecution by humans, this species is considered threatened and needs further protection to ensure its survival and the health of St. Lucian ecology.

■

■ *Enchelynassa canina*, the viper moray eel, gets its name from the long, fang-like teeth housed in strongly arched jaws where only the tips of the jaws meet. These imposing daggers are visible even when the mouth is closed and, like other morays, this species has an extra single row of teeth along the roof of the mouth. The arched jaws are perfect for wrapping around their preferred prey, other fishes, so that all the teeth gain purchase on the victim.

Viper moray eels are ambush predators that are usually seen with only their head sticking out of the reef, waiting for prey to swim by. When their meal gets close to their hiding place, the eel explodes out, grabs the fish (sometimes an octopus or squid), and then retreats back into the crevice to swallow the meal whole.

Large, thick-bodied eels, viper morays can grow up to 5 feet long. This species occurs along both Pacific coasts, from the Indo-Pacific in the west to Panama in the east. They are found on outer reefs and reef flats down to 100 feet and are thought to be nocturnal (active at night), although divers often encounter them during crepuscular (dusk and dawn) dives as well.

■

■ Also known as the canebrake, the timber rattlesnake, *Crotalus horridus*, is arguably the most widely distributed rattlesnake in North America. It occurs from Minnesota to northern Florida and from New England to Oklahoma. Timber rattlers inhabit nearly every natural habitat and appear to avoid urban areas and human development. They can be found on prairies, on mountains, and near water.

Timber rattlesnakes get up to 6 feet long and have long fangs that can swing out during a strike. Their venom is used to kill mice, rats, gophers, rabbits, birds, and lizards for food. They are ambush predators that lash out at unwary prey, quickly plunge their fangs in, and then retreat to safety while the venom does its job. Equally active day and night, they can successfully strike in total darkness because of the heat-sensing pits below their nostrils that reveal the heat signature of prey.

■

The freshwater wolf fish, *Hoplias aimara*, is a toothy predator found throughout northern South America. It grows up to 3 feet long, weighs as much as 80 pounds, and inhabits rivers and streams, where it tends to lie on the bottom using its camouflage to blend into the gravelly substrate. The genus name is derived from the Greek word *hoplon*, which means "weapon" and is based on the vicious-looking mouth.

The tubular body shape of wolf fish, coupled with their large mouth full of razor-sharp teeth, make them perfectly adapted to life as an ambush predator in water that can move quite swiftly at times. Wolf fish tend to hang out in eddies and slack water behind underwater structures. From here, they explode off the bottom after passersby and slam into them with their mouth wide open, as their jaws slam shut like a bear trap. Then they swim back to safety out of the current and reorient their prey head first to be swallowed whole.

■ Ball pythons, *Python regius*, are the most common snakes in the pet trade due to their slow growth rate, relatively small maximum size (compared to other constrictors), and, most important, their docile demeanor. They can be handled with very little risk of being bitten. Their common name arises from their tendency to coil into a ball when threatened, tucking their head and neck inside for protection. One could literally "roll" this snake when it's balled up.

Native to West Africa, ball pythons live mainly in savannas and grasslands, often frequenting termite mounds and mammal burrows. They use these underground sites as shelter from predators and to cool themselves during the hottest times of the day. They also use termite mounds as beacons for food, since nearly every small mammal in their environment also hides inside the mounds. These snakes regularly eat mice, rats, shrews, and ground squirrels and do so by striking at prey, gripping them in their mouths lined with six rows of rear-directed teeth, and then wrapping them up with life-ending coils.

■

A large species of pufferfish from the eastern Pacific and Indo-Pacific Ocean, *Arothron meleagris* is known as the guineafowl pufferfish. It lives on shallow coral and rock reefs where it feeds on a variety of prey, including sponges, bivalves, snails, crabs, and encrusting organisms. Puffers can eat hard prey because they have strong beak-like teeth, for which the group is named, and mighty jaw muscles.

Like other puffers, they possess the uncanny ability to nearly triple their girth by filling with water (or with air, if lifted out of the water). When threatened or captured by a predator, they quickly inhale water to fill their stomach. A valve inside the mouth seals the water from the inside, so it's nearly impossible to force them to expel it. Their skin is highly flexible, allowing it to stretch significantly during inflation, and is covered with spiny prickles. The increased girth makes them nearly impossible to swallow, and the predator usually spits them out to freedom.

With the evolution of the ability to fill with water came a major problem—the ribs could be broken every time they swelled, leading to punctured internal organs and death. Luckily, evolution has also led to the secondary loss of ribs in puffers to let them expand safely.

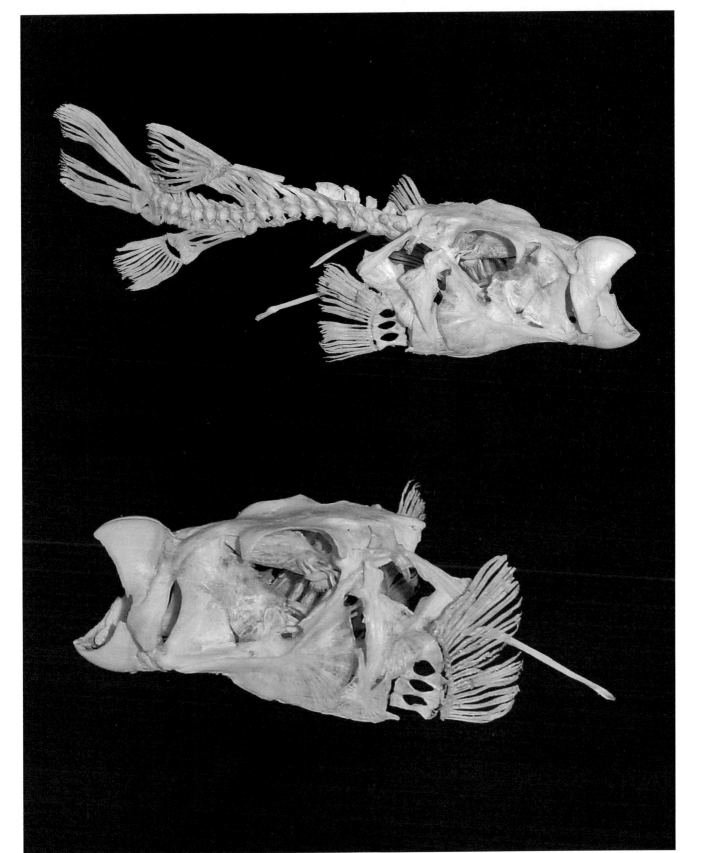

One of nature's most easily recognized and most feared snakes, the spectacled cobra *Naja naja*, also known as the Indian cobra, occurs on the Indian subcontinent and throughout much of Pakistan, Nepal, and Sri Lanka. It prefers wet habitats, such as tropical forests and paddy fields. Its pursuit of rodent prey often brings it near human agricultural lands and leads to thousands of human envenomations per year, many fatal.

Spectacled cobras are identified by the striking wire-rim eyeglass pattern on the back, and sometimes front, of their hood. This pattern is barely visible when the snake is at rest, but when fully flared, the skin stretches to reveal the mark believed to have been put there by the thumb and forefinger of Buddha. The hood is created by the actions of specialized ribs and muscles near the head. The hood ribs are much straighter than the rest and can be spread widely to expand the neck. When coupled with loud hissing and their ability to stand one-third of their 7-foot body upright, they are quite an imposing animal.

This species is quick to respond to danger and is happy to throw rapid strikes at anything posing a threat. The venom is highly neurotoxic and kills by paralyzing muscles such as the diaphragm and heart.

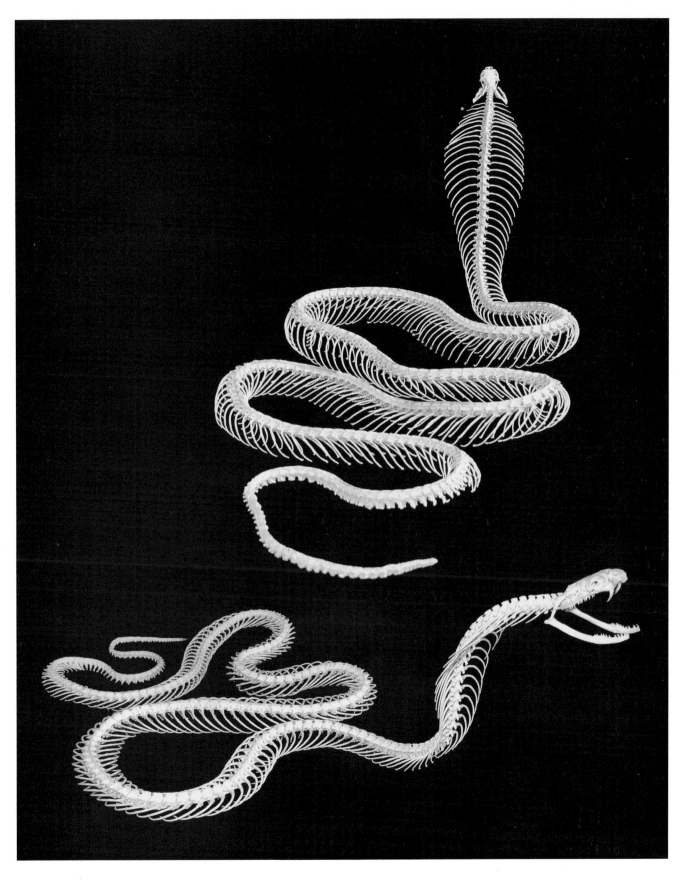

The spotted scorpionfish, *Scorpaena plumieri*, occurs in the subtropical and tropical waters of the Western Atlantic, including the Gulf of Mexico, and has a small pocket on the west coast of Africa. It is abundant on coral reefs, worm-rock reefs, rubble piles, sunken ships, and artificial reefs.

Like other scorpionfishes, spotteds spend their time hiding on the bottom, with terrific camouflage resembling a rock or other reef structure. They often get missed by waders, snorkelers, or divers descending to the ocean floor, where a poorly placed foot or hand is then pierced by the venomous dorsal spines.

Spotted scorpionfish are ambush hunters. They lie in wait for hours, sitting motionless while resembling a rock, waiting for some off-guard small fish or shrimp to venture close. A scorpionfish can explode its mouth open, shoot out its jaws, inhale a volume of water housing the prey, and close its mouth again, all in less than 20 milliseconds. The fastest human eye blink takes approximately 100 milliseconds, so these fish could theoretically feed five times in the time it takes you to blink.

Probably the most easily recognized reef fish on Earth, the yellow tang, *Zebrasoma flavescens*, is blindingly yellow. It is found on barrier and fringing reefs of Hawaii and other Pacific islands as far west as Japan. Found alone or in small, loose groups, they live in water usually less than 150 feet deep and grow up to 8 inches long.

Yellow tangs have downturned snouts tipped with scissor-like jaws that are armed with small, rake-like teeth. They swim along the reef grazing on filamentous algae, leafier seaweed, and reef-dwelling, small invertebrates such as amphipods.

This species is common prey for many other reef inhabitants, including larger fishes, octopi, and large crabs. To deter predators, yellow tangs have numerous sharp spines in their dorsal, anal, and pelvic fins that easily pierce soft flesh. Also, like other tangs and surgeonfishes, yellows possess an extremely sharp spine on each side of the base of the tail. These spines can be raised to reveal a scalpel-like edge (thus the name *surgeonfishes*), which can be used to slash and gash anything that poses a threat to survival, territory, or mates.

The pine snake, *Pituophis melanoleucas*, has a patchy distribution throughout the southeastern United States. Accomplished burrowers, they spend much of their time underground and thus live mainly in sandy or loose-soil habitats, including pine barrens, grasslands, farmlands, and some forests. They can reach 7 feet long and are powerfully built, thick-bodied snakes. As a member of the colubrid family, they kill by constricting their prey, which consists mainly of mammals such as pocket gophers, shrews, mice, and baby rabbits they find in dens.

Pine snakes are willing to stand their ground against potential threats. They will puff up (by inhaling extra air and spreading their ribs), hiss loudly, raise their head, and even shake their tails under the leaves to resemble the rattle of a rattlesnake to ward off danger. Their main predators are raptors, but they are also eaten by some mammals, such as raccoons, opossums, coyotes, and foxes.

Probably the most infamous snake in North America, the western diamondback rattlesnake, *Crotalus atrox*, has a reputation for being an ill-tempered, strike-happy pit viper. This reputation is somewhat deserved, as this snake is very prone to striking and is glad to stand its ground against threats as large as frightened horses. However, like any venomous snake, it simply wants to be left alone and poses no threat to humans, livestock, or pets if ignored.

Westerns grow up to 6 feet long and inhabit deserts, grasslands, forests, rocky slopes, and coasts of the southwestern United States up to elevations over a mile high. Their venom is used to kill mice, rats, gophers, rabbits, birds, and lizards for food. They are ambush predators that lash out at unsuspecting prey, quickly plunge their fangs in to deliver venom, and retreat to safety. Because of the heat-sensing pits below their nostrils that reveal the body heat of their prey, they can successfully strike in total darkness.

Western diamondback venom contains enzymes that degrade proteins, plus toxins that destroy blood cells, blood vessels, and muscles, including the heart. Their prey quickly dies and is partially digested by the time the snake tracks it down by trailing its scent and warm footprints through the dark.

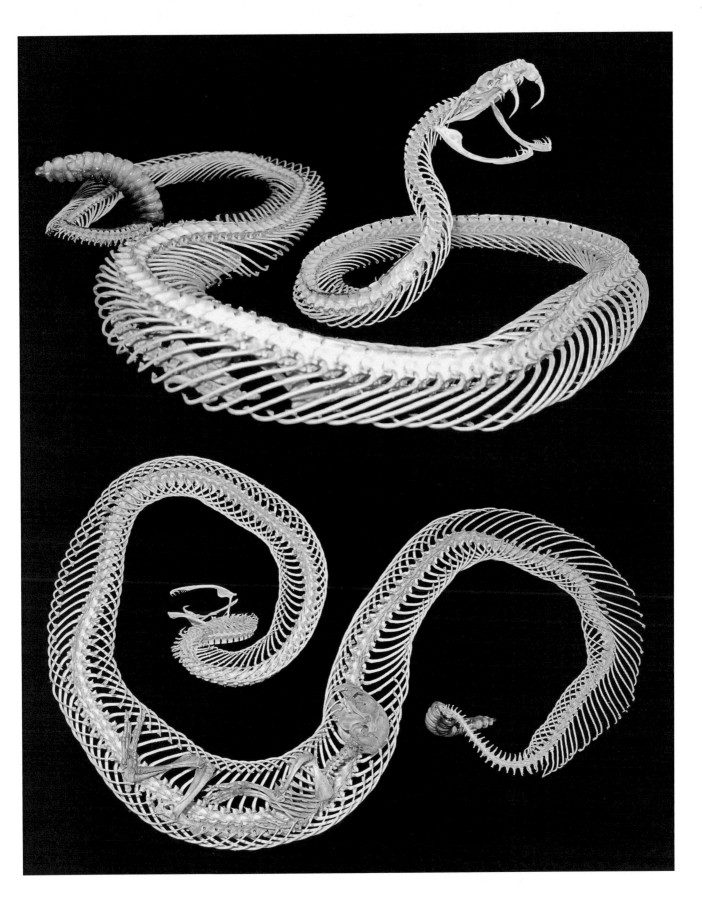

■ The white-spotted filefish, *Cantherhines dumerilii*, also known as the yelloweye filefish, has some of the most remarkable jaws in the ocean. This species only grows up to 15 inches long, but it can tear through reef structures as if it has vice grips on its snout. It eats crabs, sponges, mollusks, urchins, and most impressively, the tips of live, branching corals. The jaws are powered by enormous muscles, and the teeth and jaw bones are thick and stout, able to handle bite pressures that shatter rock. White-spotted filefish eat the entire piece of coral, digest the soft flesh off the coral skeleton, and expel large pieces of coral, like those seen opposite, collected from the gut of this individual.

Yelloweye filefish hail from the Indo-Pacific, where they inhabit coral reefs, lagoons, and outer reef edges in water up to 100 feet deep. They are highly cautious and quickly retreat into reef crevices when threatened. Once hidden, they erect their dorsal spine as an anchor into the reef structure so that predators can't pull or suck them out to be eaten.

■ The long, rear-directed teeth on both the upper and lower jaws of *Corallus caninus*, the emerald tree boa, give this species its name. These teeth are extremely useful when capturing elusive, tree-dwelling prey such as arboreal rodents and birds. Emerald tree boas have been observed snatching birds from the air as they fly by a perfectly camouflaged snake with its leaf green coloration. The snake launches an explosive attack, coils its prey in a life-ending stranglehold, and swallows it whole once it dies. The heat-sensing pits along their mouths help them see the infrared radiation generated by their warm-blooded prey, and emeralds have more heat-sensing pits than any other boa species.

Emerald tree boas are found throughout South American rainforests, and some specimens reach nearly 10 feet in length, though 6-footers are much more common. A 10-foot emerald tree boa would have nearly 100 teeth in its mouth, and some teeth would be over half an inch long.

■

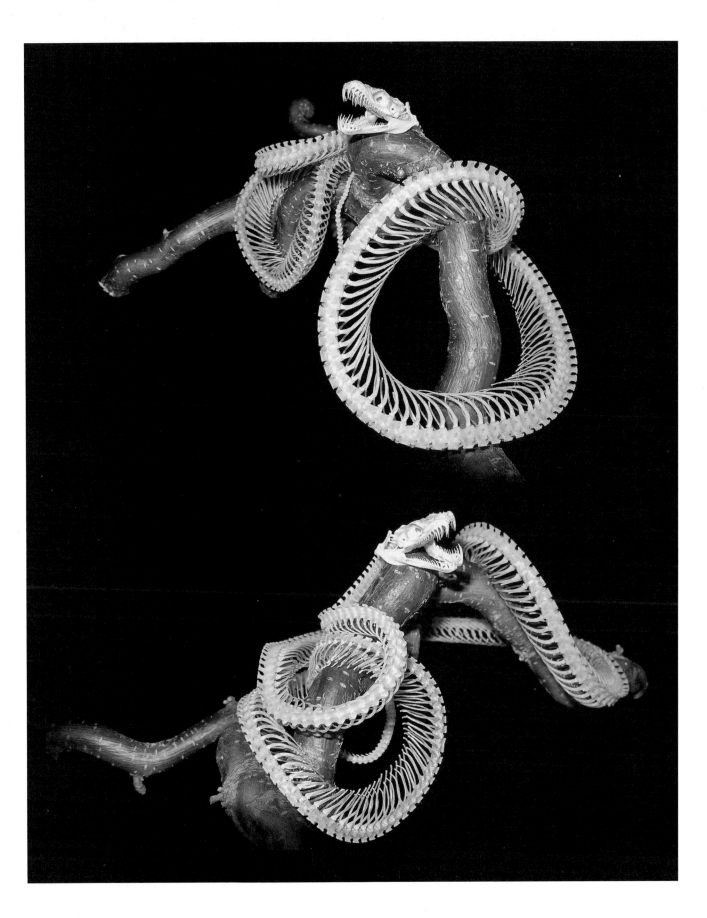

■ Bearded dragons, *Pogona vitticeps*, are found throughout Australia, where they live in nearly every habitat, including rocky crags, scrublands, deserts, forests, and beaches. Their name refers to the dark throat pigment that appears when they feel threatened by a predator or a rival. The black throat, coupled with spiny skin along the edge of the skull, creates the appearance of a "beard." A favorite of the pet trade because of their relative ease of care and friendly disposition, owners of single dragons rarely get to see the "beard" unless they trick their pet with a mirror.

Adult dragons are extremely territorial and will fight over everything from space to food to mates. If posturing doesn't work to scare off a rival, they will resort to all-out battle, where they try to push the other down and pin them, showing dominance. The winner gets to mate or gets the best territory for food, which includes leafy vegetation, flowers, insects, and worms that they process with their sharp, spade-shaped teeth.

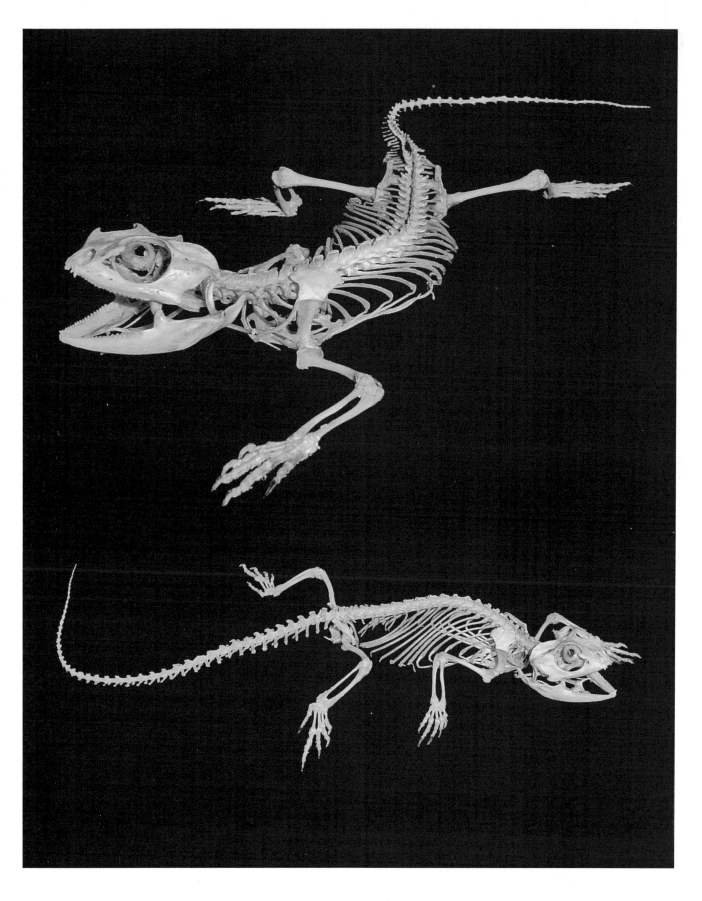

The redbreasted wrasse, *Cheilinus fasciatus*, looks more like a red-headed woodpecker than a fish, with its red head, black and white body, and pointed snout. It even uses its pectoral fins to "flap" along the reef in search of food. Redbreasted wrasses inhabit coastal lagoons, coral reefs, and rubble piles throughout the Indo-Pacific in water up to 150 feet deep. They are usually found individually but form strong pairs during the breeding season.

This species reaches 18 inches long and has specialized teeth to capture and process a diverse assortment of hard prey, such as clams, mussels, snails, crabs, and urchins. Specialized and sophisticated jaws within their throat allow them to process prey that most other fish must ignore. They pluck their prey off the bottom using their long canines and maneuver it into their throat for crunching and grinding.

A venomous pit viper, *Bothrops asper*, also known as the terciopelo (Spanish for "velvet"), ranges from southern Mexico to northern South America. This species has been called the "ultimate pit viper" by some because of its large body size, long fang length, and frequent bites on humans. It is the largest of the lanceheads, exceeding 8 feet in length, and the most commonly encountered by humans because of its vast home range and affinity for human dwellings, where it searches for rats and other small mammals.

Terciopelos are considered irritable and unpredictable and have been known to stand their ground when threatened. They can be an aggressive species if cornered and have the hardware to administer a fatal bite; their fangs can exceed 1 inch long! At times they rear up when threatened and sometimes deliver bites above the knees, though this is uncommon and can easily be avoided by leaving these snakes alone.

■ The jumping viper, *Atropoides nummifer*, gets its name from its ability to strike at distances much farther away than one would expect for such a short-bodied, heavy snake. Though they do not actually "jump" when striking, they do lunge more than half their body length and can strike quite high for a snake their size, up to 24 inches. This capacity is probably due to the disproportionately large dorsal processes off each of their vertebrae. The length of these levers, when quickly pulled upon by back muscles, lifts the snake far off the ground.

Jumping vipers are diurnal and inhabit tropical rainforests and lower cloud forests up to a mile high. They are also common inhabitants of plantations, where they pursue frogs, lizards, and rodents as prey. Unfortunately, this overlap with humans often leads to their demise, as they are feared by locals because of the painful, venomous bites they can inflict, which cause muscle and blood damage. Bites are rarely lethal if proper treatment is obtained.

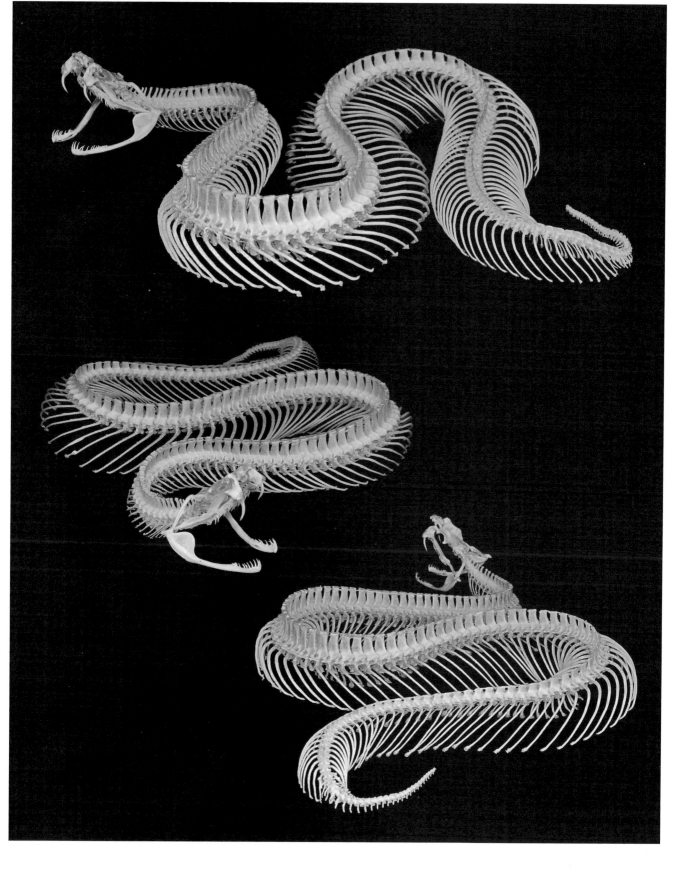

Along the continental shelf of the western Atlantic from North Carolina to Argentina lives *Mobula hypostoma*, the Atlantic devil ray. It is a pelagic ray that continuously swims through the water column in search of planktonic prey.

Atlantic devil rays are members of the class Chondrichthyes, which includes sharks, skates, rays, and chimaeras. This group is composed of animals that completely lack bones; their skeletons are made entirely of cartilage. Devil rays get their common name from the specialized, cephalic fins that border the mouth, creating the appearance of horns when viewed from directly above or below.

Like its cousin the manta ray, the Atlantic devil ray seeks out patches of plankton and swims directly through them, sometimes spiraling vertically, to engulf as much prey as it can. Evidence suggests that the cephalic fins serve to funnel prey toward the mouth as the ray propels itself through large swarms of food.

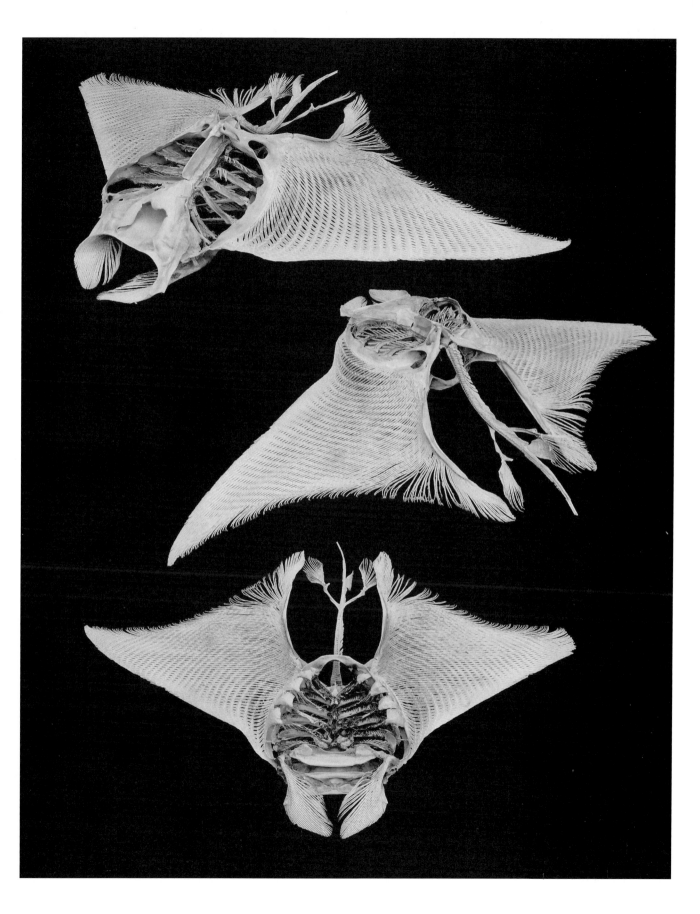

The queen parrotfish, *Scarus vetula*, is found on reefs of the Caribbean and nearby waters. It feeds on seaweed, encrusting algae, and corals, which it clips and gouges with its specialized oral jaws. The teeth on both the upper and lower jaws have become fused into a beak-like structure, driven by muscles powerful enough to take entire bites out of living corals. The food is then relocated to the throat, where a sophisticated set of pharyngeal jaws grind it into pieces that can be swallowed. Next, the organic material is extracted inside the long digestive tract and the parrotfish defecates "sand," which contributes significantly to beach material.

Queens exhibit an interesting sex strategy, common among wrasses and parrotfishes—they change sex as they age. They begin life as females, guarded within a harem, by a large male that looks completely different from the females. If he is eliminated due to predation or old age, the alpha female quickly changes sex and appearance, to assume the role of the male. Ironically, the crown-shaped coloration above the eyes is only present in mature males, so maybe they should be called "king" parrotfish.